T0243070

CAMBRIDGE LIBRARY COLLECTION

Books of enduring scholarly value

Zoology

Until the nineteenth century, the investigation of natural phenomena, plants and animals was considered either the preserve of elite scholars or a pastime for the leisured upper classes. As increasing academic rigour and systematisation was brought to the study of 'natural history', its subdisciplines were adopted into university curricula, and learned societies (such as the London Zoological Society, founded in 1826) were established to support research in these areas. These developments are reflected in the books reissued in this series, which describe the anatomy and characteristics of animals ranging from invertebrates to polar bears, fish to birds, in habitats from Arctic North America to the tropical forests of Malaysia. By the middle of the nineteenth century, this work and developments in research on fossils had resulted in the formulation of the theory of evolution.

The Natural History of the Order Cetacea

Before Henry William Dewhurst established himself in Bloomsbury, where he lectured in 1827–8 on anatomy and physiology, he served as a ship's surgeon and made a journey to Greenland and its surrounding seas in 1824. During that time he was able to study the large Arctic creatures that fascinated him, especially whales. In the decade after his return, he prepared this thorough description of polar sea life. Published in 1834, it includes many engraved illustrations. Whales were of especial interest in this period, owing to the use of their blubber in many household objects, and Dewhurst also touches on the practice of whaling. His work was one of the first studies to examine the different species of whales, as well as dolphins and other marine life. It stands as an important contribution to the development of Arctic zoology.

Cambridge University Press has long been a pioneer in the reissuing of out-of-print titles from its own backlist, producing digital reprints of books that are still sought after by scholars and students but could not be reprinted economically using traditional technology. The Cambridge Library Collection extends this activity to a wider range of books which are still of importance to researchers and professionals, either for the source material they contain, or as landmarks in the history of their academic discipline.

Drawing from the world-renowned collections in the Cambridge University Library and other partner libraries, and guided by the advice of experts in each subject area, Cambridge University Press is using state-of-the-art scanning machines in its own Printing House to capture the content of each book selected for inclusion. The files are processed to give a consistently clear, crisp image, and the books finished to the high quality standard for which the Press is recognised around the world. The latest print-on-demand technology ensures that the books will remain available indefinitely, and that orders for single or multiple copies can quickly be supplied.

The Cambridge Library Collection brings back to life books of enduring scholarly value (including out-of-copyright works originally issued by other publishers) across a wide range of disciplines in the humanities and social sciences and in science and technology.

The Natural History of the Order Cetacea

And the Oceanic Inhabitants of the Arctic Regions

H.W. DEWHURST

CAMBRIDGE
UNIVERSITY PRESS

CAMBRIDGE
UNIVERSITY PRESS

University Printing House, Cambridge, CB2 8BS, United Kingdom

Cambridge University Press is part of the University of Cambridge.
It furthers the University's mission by disseminating knowledge in the pursuit of
education, learning and research at the highest international levels of excellence.

www.cambridge.org
Information on this title: www.cambridge.org/9781108071871

© in this compilation Cambridge University Press 2014

This edition first published 1834
This digitally printed version 2014

ISBN 978-1-108-07187-1 Paperback

Drawn by R. W. Keene Esq.

A Whale Harpooned.

From a Design by Prof. Dewhurst.

Freudel. lith 24, Greek S. Soho.

THE

NATURAL HISTORY

OF THE

ORDER CETACEA,

AND THE

OCEANIC INHABITANTS

OF THE

ARCTIC REGIONS.

By HENRY WILLIAM DEWHURST, Esq.,

Surgeon-Accoucheur,

Professor of Natural History, Human, Veterinary, and Zoological Anatomy; Fellow of the West-
minster Medical, Royal Jennerian, and London Vaccine Societies; Corresponding Member of the
Worcestershire Natural History Society, Honorary Member of the London Veterinary Society;
Author of a Dictionary of Anatomy, Guide to Phrenology, Dissertation on the Component
Parts of an Animal Body, Essays on the Zoology of Man and the Study of Natural History,
Lectures on Pathological Anatomy, and the Architecture of the Human Body; Practical
Remarks on the New improved System of Warming Dwelling-Houses and Cathedrals with
Hot Water, &c.; and of numerous papers on Natural History, Mental Philosophy, &c., in
the Scientific and Literary Journals, &c.

"And ELOHIM created great *Whales*, and every living creature that moveth, which the waters brought forth abun-
dantly after their kind, and ELOHIM saw that it was good."—*Genesis* i. ver. 21.

"The works of the Lord are *wonderful* and glorious; his righteousness endureth for ever, and he hath made his
wonderful works to be remembered." Ps. cxi. 24.

"He, who does not make himself acquainted with GOD, from the consideration of Nature, will scarcely acquire
knowledge of him from any other source; for if we have not *faith* in the things which are *seen*, how shall we believe
those things which are not seen?"—*Linnæus.*

ILLUSTRATED WITH NUMEROUS LITHOGRAPHIC
AND WOOD ENGRAVINGS.

London:

PUBLISHED BY THE AUTHOR,

16, WILLIAM STREET,

WATERLOO BRIDGE ROAD.

MDCCCXXXIV.

THE

NATURAL HISTORY

OF THE

ORDER CETACEA,

AND THE

CETACEOUS INHABITANTS

OF THE

ARCTIC REGIONS

BY HENRY WILLIAM DEWHURST, ESQ.

ILLUSTRATED WITH NUMEROUS LITHOGRAPHIC
AND WOOD ENGRAVINGS.

LONDON:

PUBLISHED BY THE AUTHOR,

TO THE

Right Honorable Philip Henry.

EARL of STANHOPE.F.L.S.

PRESIDENT of the MEDICO-BOTANICAL

SOCIETY of LONDON. &c.&c.

*This Work on ARCTIC MARINE ZOOLOGY, Is Most
respectfully Dedicated as a trifling acknowledgement, of
the* CREAT BENEFITS *His* Lordship *has conferred
upon* SCIENCE, *and the zeal he has evinced for its
cultivation and Diffusion.*

By His Lordships Humble,

and Obedient Servant.

THE AUTHOR.

January, 1834.

SUBSCRIBERS.

The Royal College of Surgeons in London.
The Canterbury Literary and Philosophical Institution.

His Grace the Duke of Grafton.
His Grace the Duke of Portland.
The Most Noble the Marquis of Ailsa, K.T., F R.S.
The Most Noble the Marquis of Northampton.
The Right Hon. the Earl of Abingdon.
The Right Hon. the Earl Bathurst, K.G.
The Right Hon. the Earl of Brownlow.
The Right Hon. the Earl of Carlisle.
The Right Hon. the Earl of Cadogan.
The Right Hon. the Earl of Dartmouth.
The Righl Hon. the Earl of Denbigh.
The Right Hon. the Earl of Fitzwilliam.
The Right Hon. the Earl of Grosvenor.
The Right Hon. the Earl of Home.
The Right Hon the Earl of Lonsdale.
The Right Hon the Earl of Sheffield.
The Right Hon. the Earl of Stanhope.
The Right Hon the Earl of Tyrconnel, F.R S.
The Right Hon. the Earl of Westmorland.
The Right Hon. Lord Arden.
The Right Hon. Lord Boston.
The Right Hon. Lord Calthorpe.
The Right Hon. the Lord High Chancellor.
The Right Hon. Lord Middleton.
The Right Hon. Lord Milton.
The Right Hon. Lord Palmerston.
The Right Hon. Lord Rodney.
The Right Hon. Lord Scarsdale.
The Right Hon. Lord Sidmouth.
The Right Hon. and Most Rev. the Lord Archbishop of Cashel.
The Right Hon. and Rev. Lord Bishop of Kildare.
The Right Hon. and Rev. Lord Bishop of Bristol.
The Right Hon. and Rev. Lord Bishop of Lichfield and Coventry.
The Right Hon. and Rev. Lord Bishop of Winchester.
The Rev. and Venerable the Archdeacon of Dorset.
Rev. Professor Buckland, D.D., F.R.S., Oxford.
Rev. Professor Lee, B.D., F.R.S., Cambridge.
Rev. Professor Clark, M.D., Cambridge.
Very Rev. the Dean of Exeter and Provost of Worcester College, Oxford.
Very Rev. the Dean of Lichfield.
Very Rev. the Dean of Salisbury.
Very Rev. the Dean of Westminster.
Rev. Provost of King's College, Cambridge, 2 copies.

The Hon. and Rev. C. Bathurst, LL.D.
Hon. and Rev. F. J. Noel, A.M,
Rev. Canon Blomberg, D.D.
The Rev. Provost of Eton College.
Rev. Dr. Madan.
Rev. Dr. Faber, F.R.S.
Rev. Dr. Dick.
Rev. Dr. Gallaudet, Connecticut, United States.
Rev. Dr. Lingard, F.R.S.
Rev. Dr. Dealtry, F.R.S.
Rev. Dr. Kennedy.
Rev. Dr. Fellowes.
Rev. Dr. Nolan, F.R.S.
Rev. Dr. Rudge, F.R.S.
Rev. Canon Rogers, D.D.
Rev. Dr. Kaye.
Rev. Dr. T. Horne.
Rev. Dr. Coghlan.
Rev. Dr. Arnold.
Rev. Dr. Huddleston, F.R.S.
Rev. Mr. Hughes, A.M
Rev. Mr. Crone, A.M.
Rev. Mr. Templeman, 2 copies.
Rev. Mr. Athawes, A.M.
Rev. Mr. Goddard, A.M.
Rev. Mr. Frome, A.M.
Rev. Mr. Altham, A.M.
Rev. Mr. Brockman, A.M.
Rev. Mr. Fowle, A.M.
Rev. Mr. Bloxham, A.M.
Rev. Mr. Marsh, A.M.
Rev. William Marsh, A.M.
Rev. Mr. Townsend, A.M.
Rev. Mr. Bridges, A.M.
Rev. James Ware, A.M.
Rev. Mr. Fielden, A.M.
Rev. Mr. Bailey, A.M.
Rev. Mr. Penoyse, A.M.
Rev. Mr. Palmer, A.M.
Rev. Mr. Fergus, A.M.
Rev. Mr. Bevan, A.M., 2 copies.
Rev. Mr. Twopenny, A.M.
Rev. Charles Dewhurst, A.M.
Rev. W. Halfhead, A.M.
Rev. W. Scoresby, A.M., F.R.S.
Rev. W. James, A.M.
Rev. Mr. Methuen, M.A.
Rev. James Walker, A.M.
Rev. W. Kirby, A.M., F.R.S.
Rev. H. E. Head, A.M.
Rev. J. Slade, A.M.
Rev. Charles Day, A.M.
Rev. William Gilly, A.M.
Rev. Robert Thompson, A.M.

Rev. John Jeffery, A.M.
Rev. E. Bickerstaff, A.M.
Rev. J. Prowett, A.M.
Rev. C. Oxenden, A.M.
Rev. W. Kinsay, A.M.
Rev. W. T. Bree, A.M.
Rev. Mr. Lockwood, A.M.
Rev. William Ellis Wall, A.M.
Rev. Andrew Johnson, A.M.
Rev. S. Bird, A.M.
Rev. William Bagshaw.
Rev. Charles Simeon, A.M.
Rev. Mr. Higman, A.M., F.R.S.
Rev. Mr. Thoms, A.M.
Rev. Mr. Grove, A.M.
Rev. J. Price Jones, A.M.
Rev. T. D. Allen, A.M.
Professor Napier, F.R.S.E., Edinburgh.
Professor Coleman, F.R.S., &c.
Professor Higgins, F.R.S.E.
Professor Thompson, M.D., Glasgow.
The Right Hon. Sir Gore Ousely.
Sir Arthur De Capell Brooke, A.M., F R.S., 2 copies.
Sir Jacob Astley, M.P.
Sir William Beattie, M.D.
Sir C. Burrell, M.P.
Sir Charles Clarke, M.D., F.R.S.
Sir Andrew Halliday, M.D.K.H., F.R.S.
Sir Charles Scudamore, M.D., F.R.S.
Sir William Jardine, Bart., F.R.S.
Sir G. Whitmore, K.G.H., Col. R. E.
Sir C. Lamb.
Sir H. Halford, M.D., F.R.S., Pres. R. Coll. Phys.
Sir John Jones, Bart., Lieut.-Col. R. E., 2 copies.
Sir H. W. Wilson, 5 copies.
General Sir Herbert Taylor, G.C.H.
Sir Robert Price, Bart., M.P.
Dr. C. Holland, M.D., F.R.S.
Dr. J. E. Browne, M.D., 2 copies.
Dr. Conquest.
Dr. Harwood.
Dr. O'Beirne.
Dr. Carson.
Dr. Gilchrist.
Dr. Abercrombie.
Dr. King.
Dr. James Clark.
Dr. Yates.
Dr. Gibbens.
Dr. Hall.
Dr. Lempriere, M.D
Dr. Le Mann, M.D.
Dr. J. Kirby, LL.D.
Dr. Baron, M.D., F.R.S.
Dr. Brandreth.
Dr. J. Bulkeley, M.D.

Dr. Bevan, M.D.
Dr. Hancock, M.D.
Dr. Turton, M.D., F.R.S.
Dr. Hastings, M.D., F.R.S.
Dr. Harris, M.D.
Dr. Watson, M.D.
Dr. Macartney, M.D., F.R.S.
Dr. Whytehead, 2 copies.
Dr. Silliman, Yale College, United States.
Dr. Carter.
Hon. W. L. Bathurst, A.M.
Hon. R. Watson, M.P.
Hon. M. Elphinstone.
Hon. G. T. Hamilton.
Arthur Wainewright, Esq.
Robert Wainewright, Esq.
C. T. Abbott, Esq.
G. J. Squibb, Esq., M.R.C.S.
C. Peers, Esq.
P. H. Leathes, Esq., F.S.A.
S. Briggs, Esq.
W. H. Samwell, Esq.
Devereux Bowly, Esq.
Christopher Bowly, Esq.
Joseph Howes, Esq., F.R.G.S.
Robert Croome, Esq.
Timothy Stevens, Esq.
Joshua Brown, Esq.
George Waite, Esq., M.R.C.S.
William Spence, Esq., F.R.S., F.L..S, &c.
George Banks, Esq.
John Scott Gould, Esq.
W. Leatham, Esq.
George Silpin, Esq.
M. S. Rhodes, Esq.
W. Addison, Esq., M.R.C.S.
John St. John Long, Esq., M.R.S.L.
W. Money, Esq., M.R.C.S.
J. Stevenson, Esq., M.R.C.S.
G. Rogerson, Esq., M.R.C.S.
N. Troughton, Esq., M.R.C.S.
C. Whitlaw, Esq., M.R.C.S.
H. Thompson, Esq., M.R.C.S.
J. H. Curtis, Esq., F.R.S., M.R.C.S.
J. Forbes, Esq., A.L.S.C.M.H.S., &c.
W. H. Wiffen, Esq., F.R.S.L.
T. Castle, Esq., M.D., F.L.S., &c., Trinity College, Cambridge.
W. Swainson, Esq., F.R.S., &c.
Sharon Turner, Esq., F.S.A.
Alfred Turner, Esq., 2 copies.
H. W. Bull, Esq., R.N., M.R.C.S.L.
J. P. Stone, Esq., F.R.C.S.
W. Cunningham, Esq.
W. Chambers, Esq.
Major Richardson.
W. Holmes, Esq,, C.E.

Major Bradford.
Major Harding.
Major-General Viney, R.A, M.R.S.A., F.H.S.
Major Paine.
Captain Ross, R.N.
Captain Bayfield, R.S.M.
Captain H. W. Bayfield, R.N.
Colonel Woodroffe.
Captain Forman, R.N.
Captain Brown,F.R.S.E,
Captain Manby, F.R.S.
Alexander Rowland, Esq.
R. W. Keene, Esq.
—— Sowerby, Esq.
W. Greathead, Esq.
J. O. N. Rutter, Esq., F.R.S.
David Patterson, Esq., M.R.C, S.E.
Henry Braddon, Esq.
W. Goodwin, Esq., M.R.C.S., V.S.
W. J. A. Abington, Esq., B.A., Trinity College, Cambridge.
A. Langdon, Esq., F.S.A., Ditto.
Alexander Nash, Esq., Ditto.
J. P. Jarker, Esq.
W. Morgan, Esq., F.R.S.
C. J. Wright, Esq., A.M.
G. Warren, Esq., M.R.C.S.
J. F. Tuthill, Esq.
Chandos Leigh, Esq.
E. W. Brayley, Esq., F.S.A.
John Lewis, Esq.
R. Bevan, Esq.
Thomas Spong, Esq., Millhall.
A. Hoadley, Esq., Maidstone.
J. Priaulx, Esq.
J. Thompson, Esq.
J. Kilderson, Esq.
J. Jeffreys, Esq.
J. D. Saull, Esq.
H. H. Hoare, Esq.
J. S. Lomax, Esq.
T. H. B. Estcourt, Esq., M.P.
George Tapps, Esq., M.P.
J. S. Plumptre, Esq., M.P.
Dawson G. H. Pennant, Esq., M.P.
L. W. Dillwyn, Esq., M.P.
F. Baring, Esq., M.P.
G. Finch, Esq., M.P.
G. J. Harcourt, Esq., M.P.
H. T. Hope, Esq., M.P.
R Slaney, Esq., M.P.
J. J. Briscoe, Esq., M.P.
Benjamin Hawes, Esq., M.P.
Mr. Alderman Kelly.
Thomas Moore, Esq.

Henry Philip Hope, Esq.
James Wilson, Esq., &c., F.R.S.E., Edinburgh, 2 copies.
John Parry, Esq., A.M.
W. Crawford, Esq., 2 copies.
George Fairholme, Esq.
S. L. Francis, Esq.
Leitch Ritchie, Esq.
—— Dill, Esq., M.R.C.S.
—— Newnham, Esq.
William Marsden, Esq., F.R.S.
C. Conway, Esq.
John Nash, Esq., F.S.A.
Oswald Smith, Esq.
C. R. Cockerell, Esq.
S. E. Drax, Esq.
—— Newberry, Esq.
R. W. Blencowe, Esq.
John Mortlock, Esq.
William De Capell Brooke, Esq.
J. H. Paget, Esq., 2 copies.
George H. Fielding, Esq.
William Sells, Esq.
A. Mackenzie, Esq.
—— Taylor, Esq.
—— Lawrence, Esq.
—— Moore, Esq.
Meyrick Fuller, Esq.
R. J. Murchison, Esq., F.G S., &c.
Jonathan Green, Esq.
John Heathcote, Esq.
Hudson Gurney, Esq., V.P., F.R.S.
N. Garry, Esq.
N. Gould, Esq., F.Z.S., F.L.S., &c.
Captain Green, R.N.
Thomas Calverley, Esq.
L. Cottingham, Esq.
John.Weyland, Esq.
George Montagu, Esq., R.N.
—— Davidson, Esq,
Harry Lupton, Esq.
Henry Jackson, Esq.
Wilson Overend, Esq.
Samuel Gurney, Esq.
John Forster, Esq.
John Evans, Esq.
W. Landon, Esq.
Edwin Lees, Esq.
Thomas Fussell, Esq.
Henry Boys, Esq.
Mr. Berryman,
The Most Noble the Marchioness of Hastings.
The Right Hon the Countess of Howe.
The Right Hon. the Countess of Pomfret.
The Right Hon. the Dowager Countess of Warwick.

The Right Hon. the Countess of Surrey.
The Right Hon. Lady Noel Byron.
The Right Hon. Lady Palmer.
The Right Hon. Lady Lilford.
The Right Hon. Lady Rolle.
The Hon. Lady Brooke.
The Ladies Fitzpatrick.
The Hon. Mrs. Brooke.
The Hon. Mrs. Pelham.
Miss J. M. Currer.
Miss Masters.
Miss Sykes.
Miss Tate.
Mrs. Clark.
Mrs. Dawson.
Mrs. Thistlethwaite.
Mrs. Tibbitts.

This list is made up to the last moment of going to press.

LADIES, NOBILITY, CLERGY, AND GENTRY,

SUBSCRIBERS TO THIS WORK.

MY LORDS, LADIES, AND GENTLEMEN,

When I consider the circumstances under which I had the honour of soliciting many of you to kindly patronize this work, I feel it my bounden duty to return you my sincere thanks for your prompt compliance with my wishes. To such of you as are dignitaries and clergymen of the established church, my gratitude is in particular due, not only for the support you have individually rendered me, but likewise for the kind interest and zeal you have taken in promoting the success of this volume among your friends and acquaintance. I am aware that you may deem it invidious for me to mention any of your names who have thus benefitted me ; but I conceive that I should be guilty of an act of ingratitude were I not to mention the names of the Earls of Abingdon and Stanhope ; William Cunninghame, Esq.; Rev. Provost of King's College, Cambridge ; Rev. W. Jeffreys ; Henry Thompson, Esq., of York ; Rev. Dr. Lingard ; Thomas Castle, Esq., M.D., of Trinity College, Cambridge ; Rev. Professor Buckland, D. D., Oxford ; Sharon Turner, Esq., F.A.S.; Mrs. Arthur Wainewright; P. H. Leathes, Esq., F.A.S. ; Edwin Lees, Esq., of Worcester ; The Hon. and Rev. Charles Bathurst, LL. D., and lastly, although not the least, Sir Arthur de Capell Brooke, to whose personal kindness, and subsequent letters of recommendation, no less than upwards of seventy of the most illustrious among you, kindly consented to patronize this volume.

Having in its proper place explained the necessity of a work of this nature, I need not here repeat it ; and in conclusion have only to observe, that I earnestly hope my humble endeavours to diffuse information on the Marine Zoology of the Polar Regions, and to fill up a chasm in this interesting department of Zoological Literature, may be so fortunate as to meet your approbation: should this be the case, I may be tempted to write another volume, illustrative of the Natural History of the Quadrupeds and Birds inhabiting those bleak abodes. With every sentiment of respect and gratitude,

I have the honour
to remain,
My Lords, Ladies, and Gentlemen,
Your humble, obedient, and
obliged Servant,
H. W. DEWHURST.
January 1834.

P. S. I have to apologize for the delay that has taken place in the publication of this work, but it has occurred from circumstances over which I had no control.

PREFACE.

It has long been a matter of surprise to Naturalists that no English Zoologist has considered it worth while to delineate accurately the *Oceanic Inhabitants of the Arctic Regions*—a portion of Natural History embracing not only subjects of the highest interest to the Philosopher, Theologian, and all lovers of the works of an Omnipotent and All-bountiful Creator, but, in fact, forming the wonders of the creation, inasmuch as some of these (the WHALE *genus* for example) are the most stupendous creatures which can possibly be conceived by human beings, who breathe the vital air in common with themselves, their offspring being brought forth, and during their earlier periods of existence nourished, in a similar manner to man and the higher order of animals by which he is surrounded, but whose residence is in an element where the latter would perish, from the habitation of the former being the vast and mighty deep. Yet, the published descriptions of these creatures are either intermixed with fabulous narrations or else they are too superficial in point of matter of fact, as is the case with those related in the Natural History of Count Buffon, Dr. Goldsmith's Animated Nature, and others too numerous here to mention.

Notwithstanding so much has been said in the above-mentioned publications, yet, strange to say, there is no work extant wherein the *Zoology of the Order Cetácea, and the Oceanic Inhabitants of the Arctic Regions,* is concisely and accurately described. The author having, in order to obtain the requisite materials for this work, visited the Greenland and adjacent seas in the year 1824, as principal surgeon of the ship Neptune of London, where he had numerous opportunities of observing these creatures, of which he availed himself; and has accordingly endeavoured to fill up this important *chasm* in Zoological Literature, by correctly detailing his own observations, combined with all that was accurate and valuable in the various authors who have written upon Zoology, or scattered in the various periodicals.

This work embraces a comprehensive account of the Natural History of every known animal inhabiting the Polar Seas, from the mighty Leviathan of the deep down to the almost invisible animalculæ; and from the circumstance of these dreary regions having been so recently visited by that intrepid and scientific navigator, Captain Ross, who has so much enriched our geographical knowledge of these bleak abodes, by his valuable and important discoveries; the author flatters himself, in consequence of this fact, that this hitherto neglected department of the Arctic Oceanic Zoology, will be perused with a greater degree of interest by his subscribers, than it might otherwise have possessed. The author has endeavoured to render the language easy and comprehensive (devoid of technical terms), but at the same time equally valuable to the practical Zoologist and the man of science, as it is interesting and instructive to the general reader.

Palmam qui meruit ferat is the motto, which was

borne by the gallant Nelson, the immortal hero of the
Nile and Trafalgar; and, having had the advantage of
talented auxiliaries in the mechanical department of this
work, I conceive that it is but an act of justice I am
doing, when I state that the Frontispiece* and the
Lithographic plates were drawn by my talented and
ingenious pupil and friend, Mr. Frederick Sexton,†
whose graphic illustrations, particularly of Sacred and
Zoological Subjects, will shortly place him at the head
of his profession. I am likewise indebted to him and
my scientific friend and student, Mr. R. W. Keene, for
many of the embellishments on wood, which have been
drawn from my own designs and selections. The wood
engravings were executed by Mr. John Berryman, of
Gough Square, London. I have also to return my
thanks to Mr. Sowerby and Mr. Loudon, the celebrated
Zoologists, and to Mr. Limbird, the publisher of the
Mirror, for the loan of several illustrations; likewise to
Sir Arthur de Capell Brooke for much valuable information, which will be found mentioned in several parts of
the volume. And lastly, to G. J. Guthrie, Esq., F.R.S.,
President of, and Professor of Anatomy and Surgery to,
the Royal College of Surgeons, for many facilities I
have obtained in the course of my Zoological Studies,
and in the progress of this work, from the Hunterian
Museum of that splendid establishment.

---

* The original drawing was presented me by my friend and pupil,
R. W. Keene, Esq., who executed it from my own design, made in the
Arctic Seas in 1824.

† This gentleman is connected with the Lithographic Establishment of
Baron Friedel, in Greek Street, Soho Square, by whom the plates were
printed.

The Zoological arrangements of the *Asterias* or *Star Fishes* at page 279, should have been thus placed, and which the reader is requested to bear in mind when perusing Division IV., Radiata, that the following classification should have been placed as the head to those Astites. "Class I., Echinodermata. Order, Pedicellata. Genus, Asterias." This is in conformity with the arrangement of the Animal Kingdom by the late Baron Cuvier. The omission at the proper part of the work arose in consequence of a portion of the manuscript having been mislaid, and its loss was not discovered until too late to rectify it.

LIST OF PLATES.

CONTENTS.

NATURAL HISTORY

OF

THE ORDER CETÀCEA,

&c. &c.

" There are none, O Lord, who can compete with thee in the creation
of thy wondrous and stupendous works; for they alone are sufficient to
convince mankind of thy existence."

INTRODUCTION.

THE cetàceous animals constitute the last order of the
class *mammàlia,* or mammiférous animals, in most of
the modern systems of zoology, especially those of
Linnæus, Blumenbach, and Cuvier, by whom the natural history of these animals has been denominated
cetólogy: this may be defined to be that department
of zoology which treats of the structure, economy, and
history of cetàceous animals, or of *whales* and those inhabitants of the deep which resemble them in their
anatomical structure.

Few of these animals appear to have been known to
the ancients, as we meet with but little respecting them
in the writings of the first zoologists. Both Aristotle and
Pliny, however, mention several of those species with
which naturalists are now acquainted. Thus the former,

B

in his "Historia Animalium," lib. iii. cap. 12, speaks of the *great* or *Greenland whale,* whilst, in the same book, he treats of the *dolphin* and *porpoise.* The account which he gives us of these species is indeed very imperfect, and a good deal of the marvellous is mixed with his descriptions, but he is much more to be relied upon than any of his successors in the ancient schools of philosophy and science : in particular, his natural history of the dolphin is the most faithful of any that we find in the ancient writers, and proves that Aristotle was well acquainted with the true form and manners of the animal which he describes, either from his own observation or from that of his assistants.

The "Natural History" of the elder Pliny abounds with observations on several species of whales, especially the *great whale,* which he describes in lib. iii. cap. 37 under the name of *musculus;* the dolphin, *delphinus,* lib. ix. cap. 9 ; the porpoise, *phocæna,* in cap. 8, and the grampus, *orca,* in cap. 6 of the same book. We are by no means certain, however, that modern writers are correct in identifying the *musculus* of Pliny with the *mysticété,* by considering them as synonymous; inasmuch as he speaks of the former as preceding another species, which he calls *balæ`na,* by way of leader; and in several parts of his work he denominates the largest species of the whale *cété.* The descriptions and relations of Pliny respecting these animals are exceedingly fanciful, showing that disposition towards the marvellous for which this zoologist is so celebrated : his account of the dolphin, in particular, is little better than an entertaining collection of fables, gleaned chiefly from the poets and travellers of the time; but his account of the grampus, and of the contests between this species and the large whales, is very respectable, and tolerably authentic.

Among the earlier naturalists of modern times, many have treated more or less minutely of cetàceous animals; as Aldrovandi, in that part of his general work entitled *Cèta*; Gesner, in his work, "De Piscibus;" Johnston, in his "Historia Naturalis de Piscibus et Cètis;" and Rondèlet, in his "Histoire des Poissons." Of these, the most respectable is Rondèlet, whose work is not unfrequently quoted with approbation: he, however, does not add much to our stock of information respecting the number of species, although he mentions some, and particularly the *B. gibbar*, which were unknown to the older naturalists. The work of Aldrovandi is, perhaps, the most imperfect and inaccurate of the four: he quotes largely, and apparently with implicit reliance, from the writings of Aristotle and Pliny, and even from the fictions of the poets.

Among the naturalists of the seventeenth century, I may mention three of our countrymen, of distinguished eminence in most branches of the science,—Willoughby, Ray, and Sibbald. Mr. Willoughby's work, "De Historiæ Piscium," edited by his friend Ray, contains many valuable remarks on the cetàceous animals, more particularly the *great whale*, the *dolphin*, the *porpoise*, and the *grampus*. This learned writer appears to be the first who marked distinctly the similarity of anatomical structure in whales and quadrupeds.

Mr. Ray, in his " Synopsis Piscium," follows his predecessors in zoology in the error of including the *cetàcea* among fishes, although he seems to be among the first who doubted the propriety of such a classification. The number of species enumerated by Ray is considerable, and includes almost all that have at different times been thrown upon the coasts of our islands.

The first work of any distinguished eminence, as a
separate treatise on the cetàcea, is the "Phalainologia"
of Sir Robert Sibbald, published in Edinburgh in
1692, and in London in 1778. In this work the
author professes to describe the rare species of whales
that have been cast on the shores of Scotland, distinguishing them, according to their natural characters,
into genera and species, and adding some observations
on the nature, origin, and use of spermaceti and am-
bergris. Considering the time at which it was written,
it is a valuable work, containing accurate descriptions,
and, in general, judicious remarks : it first treats of
whales in general ; and then distinguishes them into such
as have teeth in both jaws, such as have teeth only in
the lower jaw, and such as want teeth altogether (the
proper *balæ'na*). He particularly describes the *grampus*,
the *small spermaceti whale*, or round-headed cachalot;
the *black-headed spermaceti whale*, or great-headed cachalot; the *high-finned cachalot*, of Pennant; the *common Greenland whale*; the *pike-headed whale*; and
the *round-lipped whale*.

As far as Sir Robert Sibbald depends upon his own
observation, he appears pretty correct in his descriptions; and his work must be deemed one of the best
treatises on cetàceology, and far superior to any thing
that appeared for nearly a century afterwards.

Early in the eighteenth century, Artédi, the friend
and companion of Linnæus, composed his "Synopsis
Piscium," into which he introduced the cetàceous animals as an order of fishes. He distinguishes a greater
number of species than had before been enumerated.
His specific characters are, in general, highly expressive
and very accurate, although he appears to have copied an
error from Rondèlet, in describing the grampus as hav-

ing broad serrated teeth, a mistake into which Linnæus has also fallen.

Among the last writers who have classed the *cetàcea* as an order of fishes is Mr. Pennant, who has borrowed much of his information from Sir Robert Sibbald's works, and has also gleaned freely both from the ancients and from some modern voyages of travels and histories, as Dale's "Account of Harwich;" Martin's "History of Spitzbergen;" Crantz's "Greenland;" and Borlase's "Account of Cornwall." The blunt-headed cachalot he seems to have described entirely from his own observation. He has given a figure of the animal, with its teeth.

The most complete and scientific account of the cetàcea is, however, to be found in the "Histoire Naturelle des Cetàcées" of Count La Cépède, published in Paris in 1804. John Hunter has given the best account of the anatomy and physiology of these animals;* the Abbé Bonnatérre has described in the "Encyclopédie Méthodique," in an excellent article on "cetòlogy," their natural history; but La Cépède has condensed all that was valuable on the subject, having reduced it to form and method, and improved the whole with a very scientific arrangement and animated description, though he has included many serious errors. He has distributed the thirty-four varieties or species of *cetàcea* into two orders, the toothless and the toothed: of the former he makes two tribes and eight species; of the latter, eight tribes and twenty-six species. His division of the genera is certainly more scientific than that of any of his predecessors, inasmuch as it is founded on anatomical differences; and though the generic and specific characters are often unnecessarily long, and involve circumstances

---

* *Vide* Philosophical Transactions, vol. LXXVII. part ii. p. 371.

that are implied in the preceding characters of the order
or the genus, they are more accurate and more descrip-
tive than those of any other author with whom I am
acquainted.

These are the principal writers on zoology who have
treated on cetàcea; but there are several works, on the
productions of particular countries, which contain useful
or curious information on the same subject. Of these I
shall notice a few of the most respectable, and thus conclude my historical sketch of cetological writers. Among
the earliest of them is the "History of Iceland," by
John Anderson, a German naturalist of considerable reputation: he has described several species that were but
little known before his time, particularly the *balæna
nord-caper*, or Iceland whale; the *balæna gibbosus*, or
knobbe-fish, or scrag whale; and the *balæ'noptera jubartes*, or Jupiter fish (the pike-headed whale); and he
has interspersed some amusing particulars respecting the
manners of the Icelanders, and the methods employed
by them for taking the cetàcea, though his accounts
cannot always be received with implicit reliance.

Frederick Marten, another German, published an
"Account of Spitzbergen and the neighbouring Arctic
Regions," which is frequently referred to by Pennant
and other zoological writers, particularly as containing
the best account of the B. *gibbar*, or fin-fish, and the
butskoff, or beaked whale.

About the middle of the eighteenth century, John
Egede, a Danish missionary, who had lived many years
in (Eastern) Greenland, successfully labouring for the
conversion of the natives, having acquired a thorough
acquaintance with the productions of the natives, and
their manners, and the country, published his "Description of Greenland," which was speedily translated into
English, and published in octavo, with tolerable plates.

It contains an account of the black or Greenland whale, the fin-fish, and the narwhale, but is nothing to boast of.

The "History of Greenland," by David Crantz, a German missionary of the United Brethren, was published in English in 1767. It gives the best account of the natural history of the Polar Regions. This is confined chiefly to the first volume, which contains descriptions of thirteen species of cetàceous animals: only two, however,—the white-fish and the porpoise,—are from his own observation.

In 1751, Erich Pontoppidan, Bishop of Bergen, published, in the Danish language, his "Natural History of Norway," which was translated into English, and appeared in London in 1755. The second part of this work is chiefly devoted to zoology, and contains many particulars respecting some of the *cetàcea; as the *hual fish* or great whale, the *nebbe-hual* or beaked whale, the *narwhale*, and the *porpoise*. His account of the great whale is very minute and tolerably accurate, though in many points it borders on the marvellous. His engravings are badly executed and are incorrect.

Among the British Faunæ, I may mention Dale's " History of Harwich," Borlase's " History of Cornwall," already quoted, Neill's " Tour to the Orkney and Shetland Isles," and the Rev. Dr. Fleming's " Natural History of the Shetland Islands." The former of these last two works contains the distinguishing characters of the *delphinus deductor*, or Cáaing whale; and the latter notices the several species of cetàceous animals that have appeared on the Zetland shores. Dr. Fleming has also given an excellent account of a species of narwhale in the " Memoirs of the Wernerian Natural History of Edinburgh."

In the late Baron Cuvier's great work on the animal kingdom, very little information is given on the natural history of the cetàcea, the descriptions being extremely superficial, and mostly gleaned from preceding writers. The description given by the Baron of the most important distinguishing characters of the balænóptera rorqual is incorrect, as I shall take occasion to point out more particularly in the proper place. I must not omit to mention the works of Captain Scoresby on the Arctic Regions and the whale fishery, inasmuch as they abound with valuable information on the zoology of the Arctic Seas. The principal fault of Mr. Scoresby's works is, that in many points they are too superficial in regard to some animals, and particularly the fish tribe of the Polar Regions. At the same time, naturalists are indebted to him for the best and most accurate account of the *monodon monoceros* or narwhale, as also the *medusæ*, which form no inconsiderable portion of the oceanic Arctic inhabitants: and to the works I have mentioned it is my duty candidly to state that I am indebted for much valuable information on this interesting department of natural history.

AN ICEBERG IN LATITUDE 70° 35′.

It may be necessary for me to observe that, in my descriptions of the oceanic inhabitants of the Arctic Regions, I shall arrange them in conformity with the classification of Linnæus, Count La Cépède, Baron Cuvier, Captain Scoresby, and Dr. Traill, in two parts: first, describing the *cetàcea* both in their zoological and anatomical characters; secondly, the *fish tribes;* and ultimately concluding with an account of the *crustacea, medusæ, asteriæ,* and the minutest discoverable *animalculæ.*

I therefore commence with calling the reader's attention to the *whale,* the largest of any known animal existing at the present day, and which is, to use the language of the poet Milton,

" That sea beast,
Leviathan, which God of all his works
Created hugest that swim the ocean stream."
Paradise Lost, b. i. 138.

PART I.—CLASS MAMMÀLIA.

ORDER I.—EDENTATÆ, OR TOOTHLESS CETACEA.

GENERAL HISTORY AND CHARACTER OF WHALES.

WHALES constitute a tribe of cetàceous mammiferous animals, which, from their external appearance and habits of life, in their native element, the briny deep, appear at first sight to approach so near to the fish tribe, that it is no wonder the ancient, and even, as already mentioned, modern naturalists, who were but little acquainted with the correct history or structure of these creatures, or, in fact, any of the finny race, should arrange them as appertaining to the class of fishes.

There are no fewer than seven species of the whale which strictly appertain to the genus *balæna*; and the same number of species of the *delphinus*, or dolphin tribes, which may be considered as inhabiting the Northern or Arctic Seas. As the public are frequently led to deem the latter as whales, from their inattention to the distinguishing characters of this order of cetàcea, and as the physèters, or sperm-whale tribe, are not unfrequently captured in the Arctic Seas, to complete this natural history of the cetàceous animals, as well as to please some of my zoological friends, I shall give an account of the latter in the proper place.

The whale, however, which is considered the most important in a commercial point of view, and which, in fact, constitutes the sole object of our expeditions to the Arctic Seas, is the *balæna mysticétus*, or common black Greenland whale. This species, as I shall show hereafter, is extremely valuable, on account of the abundance of oil which it produces; and, being considerably slower in its motions as well as more timid than many of the same genus which are about the same magnitude, it is consequently much easier captured.

Mr. Ray, in his " Synopsis Piscium," has a chapter headed " *Pisces sive Cetacei Belluæ Marinæ;*" and his friend, Mr. Willoughby, in his treatise " De Historia Piscium," also considered them as a species of the class of fishes, although possessing some distinguishing peculiarities of structure. The former writer, whose natural arrangement of fishes is excellent, and deserves no common praise, divides them into two principal sections; the one comprehending those that possess *lungs* for the important offices of respiration; and the second those that have *gills*, or *branchiæ*, for performing the same function : the latter division constituting fishes in the truest sense of the word.

The reasons offered by Mr. Ray for uniting these two distinct classes and species of animals are these—

1st. Because the configuration or shape of their bodies strictly agrees with that of fishes.

2dly. Because they are entirely *naked*, or covered only with a smooth skin.

3dly. Because they are *oceanic inhabitants*, and have all the actions of fishes.

Notwithstanding all these arguments by Mr. Ray,

which are not only ingenious, but true, yet Sir Charles
Linnæus, with an accuracy of anatomical discrimination
which an enlightened posterity bids fair to honour and
esteem, has very properly referred them to the class
mammàlia, a reference extremely just, though the propriety of such an arrangement does not, at first sight,
appear correct to the mere anatomist or the historian
of nature.

Mr. Pennant, in his " British Zoology," describes
these and all other cetàceous animals found upon our
coasts under the title of *cetàceous fishes*. He objects to
the scientific arrangement of Linnæus in the following
terms : " In order to preserve the chain of beings entire,
Linnæus should have made the genus of *phocæ* or seals,
together with the *trichécus rosmarus* or walrus, immediately precede the whale, those having limbs connecting
the mammàlia, or quadrupeds, with the fish ; for the
seal is, in respect to its legs, the most imperfect of the
former class ; and in the walrus all the hinder feet
coalesce, assuming the form of a broad horizontal tail."*

Mons. Bloch has very properly excluded the whales
and other cetàceous animals, with the exception of the
delphinus phoceàna, or porpoise, from his splendid work
upon fishes ; yet, strange to say, he has, in the most
inconsistent manner, included them in one of the smaller
and late editions as a seventh class, " *les cetàcées*." In
a prefatory note he informs us, however, that Linnæus
places them at the conclusion of the mammàlia, immediately after the hog tribe. But the entire class of these
animals inserted by Bloch has been supplied from the
researches of Duhamel, with the aid of the works of
Anderson, Bonnatérre, Artédi, Ray, and Belon. Baron

* " British Zoology," vol. iii.

Cuvier, in his " Règne Animal," makes it constitute a portion of the eighth order of the mammàlia, placing it after the ruminantia.*

In regard to the cetàcea, the usual arrangement of animals by the conformation of their teeth was necessarily abandoned by Linnæus, on referring to their anatomical structure; whilst, in other orders, he considered this as contributing to form a generic character.

The natural. history of the whale is an object well worthy the utmost attention of the zoologist, theologian, and philosopher. In all probability, it was this animal that gave rise to the fabulous inventions of the ancients respecting hyperborean monsters, such as the *kraaken*, said to extend many thousand feet in length, and to resemble a bank of sand, or a reef of rocks, upon the surface of the water. Such exaggerations, however, are totally unnecessary for the purpose of exciting our wonder, inasmuch as these stupendous animals are, in their own dimensions, sufficiently gigantic. The size of whales, when sufficient time has been allowed for their full developement, is truly terrific. " There is no doubt," observed the late Baron Cuvier, " that whales have been seen, at certain epochs, and in certain seas, three hundred feet in length, and weighing more than *three hundred thousand pounds*. Among the individuals of this genus met with at the present day, at a considerable distance from the Arctic Pole, there are some from seventy to a hundred and twenty feet long." †

Whatever may have been the length of these animals in former times, previous to their frequent capture on account of their oil, I must dissent from those authors who, even in the present day, declare their average

---

* *Vide* " Translation by Ed. Griffiths, Esq., and others," vol. iv. p. 475.
† Idem, vol. iv. p. 476.

length to vary from eighty to a hundred and twenty-five feet and upwards; observing, however, that aged whales, of uncommon dimensions, are occasionally captured. I cannot help coinciding with the just observations of that truly scientific clergyman, the Rev. William Scoresby, of Exeter,* which I here quote: he says, " of three hundred and twenty-two individuals, in the capture of which I had personally been concerned, no one, I believe, ever exceeded sixty feet in length; and the largest I ever measured was fifty-eight from one extremity to the other, this being the largest *I ever saw.*" An uncommonly large whale is stated (I believe by the same gentleman) to have been caught near Spitzbergen, about twenty years ago: the baleen, or, as it is improperly called, *whalebone,* measured almost fifteen feet long, but the animal was not more than seventy feet; and the longest common Greenland whale I ever heard of was one mentioned by the late Sir Charles Guisecké, who informs us that, in the spring of 1813, one was killed at Godhawn, measuring sixty-seven feet: but these are rare instances. I therefore consider that sixty feet may be the average dimensions of the larger animals of this species, and that sixty-five feet is a magnitude which but rarely occurs.

In these remarks, I wish the reader to observe that I allude only to one species of whale, the *balæ'na mysticétus*; for the other species not unfrequently exceed this in size. The *balænóptera rórqual,†* or broad-nosed

---

* This gentleman was formerly a captain in the Greenland whale fishery, as it is erroneously denominated; and his works prove him to be a man of sterling scientific attainments. I shall occasionally quote his remarks, as I proceed, under the cognomen of Captain Scoresby, by which he is more generally known.

† An engraving of this species is inserted in a subsequent part of the work, with a description of the anatomy of the skeleton.

whale, a skeleton of which was in 1831 and 1832 exhibited on the site of the late King's Mews, at Charing-cross, measured ninety-five feet in length : this magnitude is, however, but seldom attained, on account of their frequent destruction by man and their oceanic enemies.

Captain Scoresby is of opinion that the common Greenland whales are to be met with at the present day as large as at any period since the commencement of their destruction on account of the commercial value of their oil. When fully grown, they may be considered as averaging in their measurement from forty-five to sixty-five feet, and from that to seventy feet; and, in their circumference, from thirty to forty feet.*

SPECIES I.

THE BALÆ'NA MYSTICE'TUS, OR BLACK COMMON GREENLAND WHALE.

THE ancient historians, and, in particular, the Scripture zoologists, have universally considered this animal as the " *leviathan of the deep.*" It is so mentioned in the sacred volume ; and in the same view many of the poets have described it, particularly Milton, who thus observes :—

> " Here leviathan,
> Hugest of living creatures on the deep,
> Stretched like a promontory, sleeps or swims,
> And seems a moving land, and at his gills †
> Draws in, and at his trunk spouts out, a sea."

---

* " Edinburgh Philosophical Journal," No. i. p. 83.

† Had our immortal poet possessed the least acquaintance with the natural history of this animal, he would have discovered that the whale

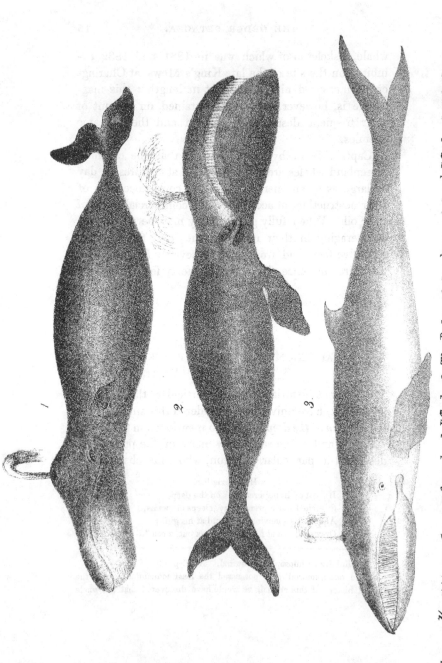

The Balæna Mysticetus, or Common Greenland Whale. 2. The Balæna Icelandica or Iceland Whale. 3. The Balænoptera Gibbar

The *balæ'na mysticétus,* or *common Greenland whale,*[*]
is supposed to be the largest animal which has yet attracted the attention of naturalists; and some of the
ancients have recorded accounts of these creatures being
upwards of nine hundred feet long. When viewed from
a distance out of the sea, it appears like a dark, confused
mass, floating just above the surface of the water;
but, when seen under favourable circumstances, we
find that this species of
whale is the thickest a little behind the fins, or swimming paws, which are in the
middle, somewhat posterior
to the articulation or joint
of the upper and lower jaws,
whence it gradually tapers
in a conical form towards
the tail, as likewise a little
towards the anterior extremity of the head. From
the neck to within about ten
or twelve feet from the tail
it is cylindrical; and beyond
this it assumes a quadrangular shape, the greatest
ridge being upwards, or on
the back, and running posteriorly nearly across the
middle of the tail.

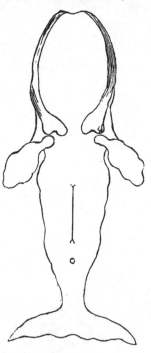

genus have *no gills,* but, as already observed, respire by lungs, in a manner
similar to mankind.

    * SYNONYMES.—*La Baleine Franche,* Bonnatérre and La Cépède ; *Com-
mon Black Whale* of Pennant and Shaw ; *Greenland Whale* of commerce ;

The head of the *balæ'na mysticétus* constitutes rather more than one-third of the entire length of the whole body, as the foregoing diagram exemplifies; but it frequently exceeds these dimensions, and approaches to one-half, and is somewhat of a triangular shape: the inferior half of the arched outline, formed by the lower jaw-bones, is flat, and measures from sixteen to twenty-five feet in length, and about ten or twelve feet in breadth. The lips extend from fifteen to twenty feet in length, and about five or six in height; forming the cavity of the mouth, they are of course attached to the lower jaw, rising from the jaw-bones at an angle of about eighty degrees, bearing the resemblance, when viewed anteriorly, of the letter U inverted. The upper jaw, including what the sailors denominate the crown-bone or skull, is bent down at the extremity, so as to completely shut the anterior and inside portions of the cavity of the mouth, being enveloped by the lips in a squamous manner at the sides.

The eyes are about the dimensions of a moderately-sized orange, being scarcely in diameter the hundredth and ninety-second part of the total length of the body, being half as large again as the visual organ of the ox; and the crystalline lens, when dried, is not much larger than a pea.* They are placed towards the posterior part of the head, it being the most convenient situation for enabling them to see both before and behind, as also to allow them to see above their head, when below the surface, where their food is principally found. The eyes

---

and the whalers; *Balæ'na Major Bippinis*, Sibbald. " Phainolog. Nov. ;" *Common Whale*, Pennant's " Brit. Zool." vol. iii.; *Great Mysticéte*, Shaw's " Zool." vol. ii. p. 2.

* *Vide* " Ency. Londin." vol. ii. p. 637; also Cuvier, " Règne Animal," vol. iv.

are guarded by eyelids and eyelashes, as in quadrupeds;
and from their actions I should conceive them to be
very quick-sighted. The seamen, in the whale-fishery,
believe them to be able to perceive objects under water
for a very considerable distance, the sense of vision
being so extremely acute; they are also of opinion that
the sight of a boat and its oars frightens them. In clear
water, they have been observed to discover one another
at an amazing distance. However, in the air they
cannot see very far, that is to say, when they are on the
surface of the water, from which cause they are consequently easily captured. Their organs are over the entrance to the ears : the functions of this latter sense are
nearly in the same perfection as the former, inasmuch
as they are warned at great distances of any danger
pressing against them. It would seem as if the Great
Author of nature had designedly given them these advantages, as they would multiply but little, in order that
they might propagate their species. It is true, however,
that they have no external organ of hearing, and the
opening leading to the internal is almost imperceptible :
were it not so, it might probably embarrass them in their
natural element : but when the thin scarf-skin is removed, a black spot is discovered behind the eye,
beneath which is the auditory canal leading to the internal organ of hearing. In short, the whale hears the
smallest sounds *under water*; but above its surface
Captain Scoresby considers it extremely dull of hearing; for a noise in the air, such as is produced by a
person shouting, is not noticed by it, even at the distance only of a ship's length; but a very slight splashing in calm water excites its attention, and occasions
great alarm.*

* The internal conformation of this organ presents some interesting

The brain of a whale, nineteen feet in length, which was examined by Captain Scoresby, weighed about three pounds and three quarters, notwithstanding the weight of the animal was near eleven thousand two hundred pounds. Here the weight of the brain was about the four thousandth part of that of the entire body;* whilst that of the brain of an adult man is about four pounds;† and, compared with that of the body, the brain forms about the thirty-third part of the weight of the whole frame.

Whales are viviparous, having but one young, and suckling it with teats, as in the other mammàlia. These organs are situated upon the abdomen; one on each side of the pudenda, near the vagina, and one about two feet apart: they appear not capable of protrusion beyond the length of a few inches. In the dead animal they are always found retracted.

The connubial intercourse of these animals is gene-

---

peculiarities, the external orifice being nothing more than a very narrow cartilaginous tube proceeding from the cavity of the tympanum, winding through a bed of fat, opening externally by a little hole, which to the eye is hardly perceptible, terminated by scarcely the vestige of a conch. This canal pierces the upper maxillary or jaw-bone, and terminates above the spiracle in an orifice, rendered by means of a small valve impenetrable to water.

The internal ear is composed of a labyrinth, a cochlea, cochlearian orifice, three semi-circular canals, a vestibulum and its orifice, and a tympanum and its membrane; also articulated osselets placed within the tympanum from its membrane to the vestibulary orifice, an eustachian tube, and a canal leading from the membrane of the tympanum, opening to the small external orifice already mentioned.—CUVIER. "Règne Animal," vol. iv. p. 414.—See also JOHN HUNTER's "Remarks on the Structure and Œconomy of Whales."

* Blumenbach's "Manual of Comp. Anatomy," translated by Lawrence and Coulson.—Edit. 1827, p. 213.

† Dewhurst's "Dissertation on the Component Parts of an Animal Body," page 57, 3d edition.

rally noticed to take place towards the latter end of summer; females with cubs, or, as the sailors call them,
suckers, along with them, being most commonly met
with in the spring of the year. When they copulate,
the female joins with the male, it is asserted, *more
humano*; and once in two years feels the accesses of
desire. Their fidelity to each other exceeds even the
constancy of birds. Captain Anderson informs us that
having struck one of two whales, a male and a female, that
were in company together, the wounded one made a long
and terrible resistance: it struck down a boat with *five*
men in it, by a single blow of the tail, by which means
all went to the bottom. The other still attended its
companion, and lent it every assistance; until, at last,
the whale that had been struck sunk under its wounds;
whilst its faithful associate, disdaining to survive the
loss, with great bellowing, stretched itself upon the dead
animal, sharing its fate.

The embryo whale, when first perceptible, is generally
about seventeen inches in length, and of a white colour;
but the cub, when born, is black, and varies from ten
to fourteen feet. Baron Cuvier has asserted it to be
twenty feet.

One cub is generally produced, occasionally two, but
never more. The maternal affection of the whale, which,
in other respects, is apparently stupid, is here striking
and interesting. The young one, being insensible to
danger, is easily harpooned, when the tender attachment
of the mother is so manifested as not unfrequently to
bring it within the reach of the whalers : hence, although
a cub is of little value, seldom producing above a ton of
oil, and frequently less than this, it is frequently struck
as a snare for its mother. In this case she joins it under
the surface of the water, whenever it has occasion for

respiration; encourages it to swim off; assists its flight, by taking it under her fin; and seldom deserts it whilst life remains. Then she is dangerous to approach, but affords frequent opportunities for attack. She loses all regard for her own personal safety, in the anxiety for the preservation of her young; dashes through the midst of her enemies; despises her threatening danger; and even voluntarily remains with her offspring, after various attacks are made upon herself by the seaman's harpoon, thus proving the truth of the ancient riddle's explanation, that " *a mother's love for her offspring enables her to brave every danger for their support and protection.*"

Captain Scoresby informs us, that in June, 1811, one of his harpooners struck a *sucker* with the hope of capturing the mother: she presently arose close to the "*fast-boat,*" and seizing the young one, dragged about one hundred fathoms of line out of the boat with considerable velocity. Again she arose to the surface, furiously darted to and fro, frequently stopping short; or suddenly changed her direction, exhibiting every possible symptom of extreme agony. For a considerable length of time she continued thus to act, although closely pursued by the boats, inspired with courage and resolution; but her concern for her offspring made her regardless of the danger by which she was surrounded. At length, one of the boats approached so near that the harpooner was enabled to heave his harpoon at her; it hit, but did not attach itself. A second harpoon was struck: this also failed to penetrate; but a third was more effectually held. However, she did not attempt to escape, notwithstanding her own sufferings, still clinging to her offspring; for she allowed the other boats to approach, so that in a few minutes three more harpoons were fastened, and in the course of an hour both were killed.

There is something extremely painful in the destruction of a whale, when thus evincing a degree of affectionate regard for its young, that would do honour to the superior intelligence of human beings; yet the objects of the adventure, the value of the prize, and the joy of the seamen with the capture, cannot be sacrificed in reflecting to the refined feelings of compassion.

When the female suckles her offspring, she throws herself upon one side, on the surface of the sea, and the young whale attaches itself to the teat. They continue at the breast for a year, during which time the sailors denominate them *short-heads.* At this time they are extremely fat, and will yield above fifty barrels of blubber, the mother being at the same time equally lean and emaciated. At the age of two years they are called *stunts,* as they do not thrive much immediately after quitting the breast; they then yield scarcely above twenty or twenty-four barrels of blubber. From that time forward they receive the appellation of *skull-fish,* and their age is wholly unknown. In the latter end of April, 1811, a sucker was captured by one of the Hull ships, having the *funis umbilicalis,* or naval-string, still attached.

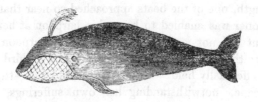

A CUB WHALE, SIXTEEN FEET LONG.

The young one continues under maternal protection for probably a year or more, or until, by the evolution of the whalebone, it is enabled to procure its own nourishment. The notches in this substance (the baleen) have been supposed a criterion of age; by which, if correct, it would appear that the whale reaches the magnitude called by the whalers *size* (*i. e.* with a six feet of whalebone) in twelve years, and attains its full growth at the age of twenty or twenty-five.

Every species of the whale genus propagates with only those of its own species; they are, however, occasionally seen in shoals of different kinds together, and make their migration, in large companies, from one ocean to another.

Although these animals, to a certain extent, are gregarious, yet, generally speaking, the common whales are solitary, travelling in pairs, male and female, unless they are attracted to a particular spot, either by an abundance of choice palatable food, or a good situation of the ice; and then they are not unfrequently found in great numbers together.

There is an interesting fact connected with the food of this species of the whale genus. They are, as already stated, inhabitants of the *olive-green waters* of the Greenland Seas, on account of the incalculable number of the *medusæ* and *animalculæ*, which occupy a *fourth* part of that ocean, or about twenty thousand square miles. These whales, from the structure of the œsophagus or gullet, which is so small as not to admit even a small herring, are unable to derive any other subsistence from the larger inhabitants of these seas. A preparation of a portion of the œsophagus or gullet of a whale is preserved in the Museum of the Royal College of Surgeons, in London.

The animalculæ forming the food of the whale consist of various species of *actiniæ, cliones, sepiæ, medusæ,
cancri,* and *helicæ;* or, at least, some of their genera
are always found wherever any tribes of whales are seen
stationary and feeding. Captain Scoresby, however,
found in the dead animals, when he was enabled to open
their stomachs, that a few *squillæ,* or shrimps, were the
only substances he discovered. In the mouth of a
whale he had just killed, he once found a quantity of this
species of the shrimp tribe. This gentleman estimates
that two square miles of the Greenland Ocean contains
23,888,000,000,000,000 of these minute creatures; and,
as this number is beyond the range of human words and
conception, he illustrates it by observing, that 80,000
persons would have been employed since the creation of
the world in counting them. He is also of opinion, " that
the *medusæ,* and other minute animals, give the peculiar
colour to the sea which is observed to be so prevalent
in the Arctic Regions; and that, from their profusion,
they are at the same time the occasion of that great
diminution of transparency which always accompanies
the olive-green colour in the blue water. Where few of
the little *medusæ* exist, the sea is uncommonly transparent. Captain Wood, when attempting the discovery of a north-east passage, in the year 1676,
sounded near Nova Zembla in eighty fathoms water,
where the bottom was not only to be seen, but even
shells lying on the ground were clearly visible."*

Buffon is even of opinion that light is capable of
penetrating six hundred feet, or 100 fathoms, into the
waters of the ocean.

---

* CLEARNESS OF THE SEA AT THE ARCTIC OCEAN.—In allusion to the
colour of the Northern Sea, the following pleasing description of its ap-

The opening of the whale's mouth, laterally is serpentine: the lips are about twenty or twenty-two feet long, and display, when open, a cavity sufficiently large to afford a reception to a ship's large jolly-boat and her crew. H. L. Duhamel-du-Monçeau, the celebrated chemist, relates, that a whale which was captured in the Bay of Sonsure, in 1726, seventy feet long, had a mouth so

---

pearance, as related by Sir Arthur de Capell Brooke, in his "Travels to the North Cape," may not be deemed uninteresting by the reader; he observes, "Nothing can be more surprising and beautiful than the singular clearness of the water of the Northern Seas. As we passed slowly over the surface, the bottom, which was in general a white sand, was clearly visible, with its minutest objects, where the depth was from twenty to twenty-five fathoms. During the whole course of the tour I made, nothing appeared to me so extraordinary as the inmost recesses of the deep thus unveiled to the eye. The surface of the ocean was unruffled by the slightest breeze, and the gentle splashing of the oars scarcely disturbed it. Hanging over the gunwale of the boat with wonder and delight, I gazed on the slowly moving scene below. Where the bottom was sandy, the different kinds of *Asteriæ* and *Echini*, and even the smallest shells, appeared at that great depth conspicuous to the eye; and the water seemed in some measure to have the effect of a magnifier, by enlarging the objects, like a telescope, and bringing them nearer. Now creeping along, we saw, far beneath, the rugged sides of a mountain rising towards our boat, the base of which perhaps was hidden some miles in the great depth below. Though moving on a level surface, it seemed almost as if we were ascending the height under us; and when we passed over its summit, which rose in appearance to within a few feet of our boat, and came again to the descent, which on this side was suddenly perpendicular, and overlooking a watery gulf, as we pushed gently over the last part of it, it seemed almost as if we had thrown ourselves down this precipice; the illusion, from the crystal clearness of the deep, actually producing a sudden start. Now we came again to a plain, and passed slowly over the submarine forests and meadows which appeared in the expanse below, inhabited, doubtless, by thousands of animals unknown to man; and I could sometimes observe large fishes of a singular shape gliding softly through the watery thickets, unconscious of what was moving above them. As we proceeded, the bottom became no longer visible; its fairy scenes gradually faded to the view, and were lost in the dark green depths of the ocean." Page 197.

wide, that when opened two men might enter it without stooping.

Both jaws are completely divested of teeth, the place of which is supplied, in the upper, by two rows of *laminæ*, denominated " *baleen*," or " *whalebone*," or, as Captain Scoresby calls them, " *fins*," which are suspended from the sides of the crown-bone. Each of these is composed of a number of stiff hairs, or bristles, placed longitudinally, side by side. They are united together, and connected to the upper part of the mouth, or crown-bone, by a species of rabbet, with a peculiar glutinous substance, which is denominated the *gum:* this is white, fibrous, tender, and tasteless, cutting like cheese, and bearing some resemblance to the kernel of a cocoa-nut. When dried, it produces on each piece of baleen a smooth and shining surface, like scales of horn.

These whalebones, or baleen, taken separately, are of an elongated form, but curved a little in their length, like a scythe, diminishing insensibly in elevation and thickness from base to point. (See *fig.* 1, p. 29.) Their edge, which is trenchant on the inferior side, is a little concave. They are furnished from bottom to top with a species of disunited bristles, forming a kind of hinge, which is more tufted and longer the more it approaches the extremity of the whalebone. At the extremity of each series, it is curved and flattened down, so as to present a smooth surface to the lips. In some whales, a curious hollow on one side, with a ridge on the other, occurs in many of the central blades of baleen, at regular intervals of six or seven inches, but for what purpose we are unacquainted.

The general colour of these horny laminæ is a blueish

black, marbled with shades of less deeper hue, and assumes somewhat of a greyish tint. In some animals it is striped longitudinally with white, and when newly cleaned, affords a fine display of colour. Sometimes they are concealed within a greyish epidermis, and then they assume that colour. Along the bone forming the superior part of the mouth, or upper jaw, the laminæ are placed with a trifling inclination from the front to the rear; the base of them entering the gum, which it traverses, and penetrates even into the jawbone, whilst the convex portion of each lamina is applied against the vault of the palate, which then appears as if bristling with very hard hairs, and the length of which, in passing the lips, constitutes a species of beard, which denomination is frequently given to them.

The palate of the whale being oval, we can easily conceive that the longest laminæ must be nearer its greatest diameter, and that the shortest must necessarily be situated near the entrance to the throat, and towards the end of the muzzle.

Some of these laminæ are twenty-five feet in length; their base, which penetrates into the gum to the depth of two or four feet, is a foot or a foot and a half in thickness; and on each side there are from three to four hundred of these laminæ: in a very small whale, the number was either eight hundred and sixteen, or eight hundred and twenty.

The whale-seamen consider that, if the largest lamina measures six feet and one inch, it is a full-sized whale: by this they are entitled to a certain sum, if it measures that length; but not if less. However, it not unfrequently measures fifteen feet in the larger whales, and from ten to twelve in those of a moderate size. The greatest breadth, which is at the gum, is about ten or twelve inches; and their arrangement bears some resemblance to a frame of saws in a saw-mill. (See *Fig. 2*, p. 30.)

Besides the laminæ I have noticed, there are other laminæ of a similar description, situated under the extremity of the palatal bones; but these are very small when compared with the preceding, they are couched one upon the other, much in the same style as the scales on the bodies of the majority of fishes. The use of these laminæ* is to prevent the extremity of the jaw, which is slender and trenchant, from wounding the upper lip.

Each of these elastic laminæ is accompanied in its developement by others, which are denominated intermediate, because they are, in fact, placed under the larger laminæ, which they separate one from the other. These are formed at the same time with the others, making but one and the same body with them, and must necessarily strengthen and maintain them in their place. In their composition, they bear a great affinity to horn. The quantity of whalebone afforded by a large whale will amount to about a ton and a half. If the "*sample-blade*," that is to say, the longest lamina of the whole series, weighs seven pounds, the whole produce may be

---

* " The plates of baleen strain the water, which the whale takes into its mouth, and retain the small animals on which it subsists. For this purpose the baleen is in sub-triangular plates, with the free edge fringed towards the mouth, the fixed edge attached to the palate, the broad end fixed to the gum, and the apex to the inside arch. These plates are placed across each other at regular distances." Rev. Dr. Fleming's " Philosophy of Zoology."

estimated at a ton, and so on in proportion. The annexed cuts will give some idea of these laminæ.

Fig. 1.

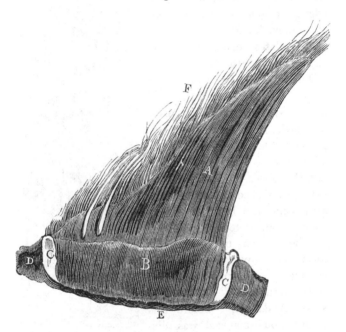

A SIDE VIEW OF ONE OF THE PLATES OF BALEEN,
OR WHALEBONE.

A, The part projecting beyond the gum. B, the portion sunk into it. C C, a white substance surrounding the whalebone, there forming a projecting bend, and also passing between the plates, constituting their external lamellæ. D D, the part analogous to the gum. E, a fleshy substance, covering the jaw-bone, and on which the inner lamella of the plate is formed. F, the plate terminating in a species of bristle or hair.

Fig. 2 represents a perpendicular section of several plates
of whalebone in their natural
situation in the gum; their inner
edges, or shortest terminations,
are removed, and the cut edges
of the plates seen from the inside of the mouth.

Fig. 2.

The upper part shows the
rough surface formed by the
hairy termination of each plate
of baleen ; the middle portion represents the distance the plates
of baleen are from each other ;
and the lower part exhibits the
white substance in which they
grow, and the basis on which
they are supported.

The baleen or whalebone,
forms an important article of
commerce. " How the ladies'
stays were made," gravely observes Anderson, " before this
commodious material was dis-
covered, history does not inform us; probably simple
slips of cane, or of some tough and pliant wood, might
have been in use before." However this may be, this
substance, after its introduction into this country, very
soon became the principal material employed in the
manufacture of that injurious article of female apparel,
the ' stays,' and likewise the hoop-petticoat, so prevalent
during the greater portion of the last century ; and which
the immortal Pope characterizes as that ' seven-fold fence,'

' Stiff with hoops, and arm'd with ribs of whale.' "

The consumption of this article was so great for some time, that the Dutch merchants are said to have received annually, from England alone, no less than £100,000. At that period, it averaged £700 per ton, which is about four times as much as it now produces, and more than eight times its worth a few years ago.

The tongue of the *balæ'na mysticètus*, or common whale, is usually thick, fleshy, fat, soft, and spongy. It occupies a large proportion of the whole cavity of the mouth, and the area formed by the whalebone. It is incapable of protrusion, being fixed from the roof to the tip, to the part extending between the two lower jaw-bones. There is, however, a little motion; some of the whale species can raise, swell, and extend it to the end of the muzzle. The base of this organ, in large whales, is covered by a flabby skin, extending towards the root, over the orifice of the œsophagus or gullet, thus rendering the entrance there so narrow, in this species, that *fish of even a moderate size cannot pass down.*

The tongue is sometimes twenty-seven feet long and from nine to twelve feet wide; but the largest I ever saw, only measured nineteen feet three inches. This organ will generally produce about six tons of oil. In some species of the whale tribe, it is covered with a slender and smooth skin, whilst in others it is altogether rough and bristly. Its colour is generally white, with small blackish spots upon the sides.

The milk of the whale was tasted by the late Dr. Jenner: according to him, it resembles that of most quadrupeds in its appearance; and it is said to be exceedingly rich and well flavoured, but containing more cream, with a greater quantity of nutritive matter.

When a whale feeds, it swims with considerable velocity below the surface of the sea, with its jaws very widely extended. A stream of water consequently enters its capacious mouth, and along with it large quantities of water-insects, &c.; the water escapes again at the sides, but the food becomes entangled, and sifted as it were, by the whalebone or baleen; which, from its compact arrangement, and the thick internal covering of hair, does not permit a single particle of the size of the smallest grain to escape.

The whales have no voice, but in the act of respiration, or blowing, as the seamen term it, they make a very loud noise, which may be heard for several miles round, and is oftentimes, nay generally, a signal to the whalers of this animal being in their vicinity. The watery vapour they discharge is ejected in a somewhat radiated form to the height of some yards, and at a distance it appears like a puff of smoke. When the animal is wounded, it is often stained with blood, and on the approach of death, jets of blood are sometimes discharged from them alone. They *blow* strongest, densest, and loudest when " running;" and in a state of alarm, or when they first appear at the surface after having been a long time down. They respire or breathe about four or five times in a minute.

Being considerably higher than the medium in which it swims, the whale can remain at the surface of the sea, with its " crown," in which the spiracles, or blow-holes, are situated, and a considerable extent of the back, above water, without any effort or motion. However, great exertion is required for its descent. The proportion of the whale appearing above water, either when alive or recently killed, is not, probably, more than the twentieth part of the whole bulk of the animal; but after death, even within a day when the putrefactive process of decomposition has commenced, the whale swells to an enormous size, until at least one-third of the carcass appears above water, and sometimes the body bursts with the force of the air generated within.

Formerly it was believed that the whales were preserved, after birth, under water, in consequence of the existence of the *oval foramen*, which in the *mammiferæ* is open previously to the birth of any viviparous animal, by which means the blood is enabled to pass from one part of the heart to another, without first circulating through the lungs. This opinion, however, is erroneous; for it is now ascertained that all the *cetàcea* can remain under water but a very short time, and are compelled to come forward frequently to the surface to respire atmospheric air through their spiracles.

These spiracles consist of two canals, situated towards the middle of the great vault of the head; a little behind is a lump or protuberance raised, upon which the orifices are situated. They proceed from the bottom of the mouth, traversing obliquely, and in a curved direction, the interior of the head, terminating towards the middle of the superior portion. They have not the same form and situation in all the *balæ'na*: in some, they have the form of two crescents, with the convexities opposed and a little separated one from the other; in others, there are two apertures, completely circular, sometimes considerably remote from each other, and sometimes so near that they appear to form but one and the same orifice, the external diameter of which constitutes about a hundredth of the total length of the animal.

The spiracles answer the purpose of expelling the water which penetrates the interior of the whale's throat, and prevents its entering the larynx, allowing the air that is necessary for the respiration of the animal to enter the lungs. They send forth so considerable a volume of water that it will instantly fill a boat. It is likewise projected with an amazing rapidity, especially when the whale is agitated by any violent feeling, and the noise is tremendous to persons unaccustomed to it. It has been stated, by some writers, that those cascades are sometimes thrown to the height of thirty or forty feet, communicating a motion to the surface of the sea which is perceptible at a distance of six thousand feet.

The organ by which the whale thus expels the water through his spiracles consists of two large membranous pouches, that lie imbedded beneath the skin, in front of the orifices with which they communicate. Some very strong fleshy fibres proceed from the circumference of the skull, uniting above these pouches, and compressing them violently at the will of the animal. When he wants to get rid of the superfluous quantity of water within his mouth, he swallows it, but at the same time closes his pharynx, or the opening of his gullet, and forces the fluid to ascend through the spiracles, where he raises, by the movement, or impresses on it, a fleshy valve placed in the spiracle itself, towards its superior extremity, below the pouches. The water then penetrates into those pouches, the valve closing, and preventing the water re-entering the mouth; and the animal, compressing the pouches with violence, ejects the water to a height proportioned to the force of the compression.*

Bulky as the whale is, and inactive, or indeed clumsy as it appears to be, one might, at first sight, suppose that all its motions would be sluggish, and its greatest exertions productive of no celerity. The fact is, however, the reverse: a whale extended motionless at the

---

* Cuvier, Règne Animal, vol. iv. p. 481.

surface of the sea, can sink in the space of five or six
seconds, or less, beyond the reach of all human enemies.
Its velocity along the surface, perpendicularly, obliquely,
or downward, is the same. Hence, for the space of a
few minutes, they are capable of darting through the
water with a velocity equal to that of the fastest ship
under sail, and of ascending with such rapidity, as to
leap entirely out of the water: this fete they sometimes
perform as an amusement apparently, to the high admiration of the distant spectator, but to the no small
terror of the inexperienced whaler. Sometimes the
whales throw themselves into a perpendicular posture,
with their heads downwards, rearing their tails high aloft
in the air, beating the water with awful violence. In
both these cases, the sea is thrown into foam, and the air
filled with vapours: the noise, in calm weather, is heard
to a great distance; and the concentric waves, produced
by the disturbance on the water, are communicated to a
considerable extent. Sometimes the whale shakes its
tremendous tail in the air, crackling like a whip, re-
sounding to the distance of many miles. When it re-
tires from the surface, it first lifts its head, then plunging
it under water, elevates its back like the segment of a
spear, deliberately rounds it away towards the extremity,
throws its tail out of water, and then disappears.

Whales descend to a great depth in the sea; some
say a mile. Captain Scoresby says he harpooned one
that descended four hundred fathoms, at the rate of
eight miles per hour; and there are instances of these
animals having been drawn up by the attached line from
a depth of seven hundred or eight hundred fathoms (one
occurred in our voyage) and sometimes the jaw and
crown bones have been found broken by the blow struck
against the bottom. Many whalers are of opinion that

whales can remain under a field of ice, or at the bottom of the sea, in shallow water, when undisturbed, for many hours at a time: whether this is the case is uncertain; but whales are seldom found sleeping; yet in calm weather, among ice, instances occasionally occur.

Mr. O'Reilly, however, was informed by a respectable Captain of a Davis's Straits' ship, that (some years previous to the year 1817, when that gentleman was in Greenland,) a native paddled alongside, making anxious expressions of useful information which he had to communicate: it was, that his companions had during three days previously observed a large whale sleeping at the bottom of a neighbouring creek. On sending some boats to the spot, and splicing together some oars, by this means sending down a harpoon, the animal was struck, and subsequently taken. Here is a curious physiological question to solve, Does the action of the lungs in this case remain suspended? or does the arterial circulation proceed so as to supply a sufficient vitality?

When I was in Greenland, in 1824, I noticed that the arterial blood was apparently of a much higher temperature in all the animals we captured than in England; and, on referring to Mr. O'Reilly's work, I find he corroborates my statement. What the temperature is I am unable to tell, our thermometer having been accidentally broken during our passage out.

The male organ of generation is a large flexible member, concealed within a longitudinal groove, the external opening of which varies from two to three feet in length. This member in the dead animal is from eight to ten feet in length, and about six inches in diameter at its root. It tapers to a point, and is perforated by the urethra; it has only one corpus cavernosum. The vent or anus is about six inches behind the pudendum of the female; but, in the male, it is more distant from the organ of generation.

COLOUR OF THE SKIN.

The colour of the balæna mysticètus is a velvet black, grey (composed of dots of blackish-brown, on a white ground), and white, with a tinge of yellow. The posterior part of the upper jaw, and part of the lower jaw, together with the fins and tail, are black. The tongue, the anterior part of the under jaw and lip; sometimes, a little of the upper jaw, at the extremities, and a portion of the belly, are white, as the annexed cut exemplifies.

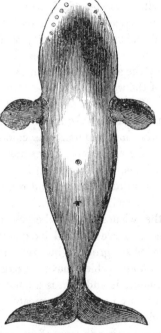

The eye-lids, the junction of the tail with the body, a portion of the axilla or arm-pits of the fins, are grey. Captain Scoresby and Mr. O'Reilly state, that they have seen piedbald whales, and one our seamen captured was of this appearance. The older animals contain the most grey and white; under-sized whales are altogether of a bluish-black, and suckers are of a pale blue or bluish-grey colour.

The skin of the body is slightly furrowed, like the water-lines in coarse laid paper; on the tail, fins, &c.,

it is smooth. The cuticle, or that external part of the skin which can be pulled off in sheets after it has been dried a little in the air, or particularly in the frost, is not thicker than parchment. The skin is very strong, although penetrated by a multitude of pores. In certain species it is more than eight inches thick.

The epidermis is very smooth, porous, and composed of several layers. It shines, because it is penetrated with a species of oil, which in the rays of the sun give it the appearance of polished metal. This oil, besides diminishing the rigidity of the skin, preserves it from those injurious changes it might otherwise experience by the alternate sojourn of the animal below the surface of the water.

The mucous tissue, or *rete mucosum*, which separates the epidermis from the true skin, is thicker than in the other mammiferæ. Its colour, which is communicated to the epidermis, varies much, not only in the different species, but likewise in individuals of the same kind, by reason of age, sex, and probably by the temperature of the usual habitat. Baron Cuvier states that almost all the whales of Spitzbergen are entirely white. Some cetàcea are marked irregularly with white upon a grey or black ground, but this appears to be the result of wounds, which have cicatrized. The flesh below the epidermis and skin is a dark red, very coarse, hard, and dry, by no means agreeable to taste, and impregnated with an unpleasant odour. Between the flesh and skin is sometimes a coat of fat, more than a foot in thickness on the head and neck. A part of this fat is so liquid that it often forms an oil without the necessity of having recourse to the process of expression, or the application of heat.

QUANTITY OF BLOOD IN A WHALE.

The quantity of blood which circulates in the whale is much greater in proportion than that which flows in the veins of quadrupeds. The diameter of the aorta, or large artery arising from the heart, is sometimes more than thirteen inches, and the late Mr. John Hunter estimated the quantity thrown into it, at every contraction of the heart, to vary from *ten* to *fifteen gallons*, and that with an immense velocity. The heart of the whale is broad and flattened, and larger in this animal, in proportion to their size, than in any quadruped, as also the blood-vessels, and particularly the veins.* The whale has a very voluminous liver, a spleen of no very great extent, a pancreas or sweet-bread very long, a bladder of middling size and elongated form. The stomach of the whale is peculiarly conformed; instead of four cavities, as in the ruminantia, there are five which are very distinct and separated from each other.

SENSE OF SMELL.

Whales are supposed to possess the organ of smell, and Count La Cépède relates the following anecdote in support of this idea. " The Vice-Admiral Pleville-le-Peley, being one day at sea with his fishers, perceived some whales above the horizon. He prepared to give way to them, but in order to stow away the quantity of cod which was in the boat, and having in the hold a great quantity of stinking and putrid water, Pleville-le-Peley ordered this pestiferous fluid to be flung into the sea. The whales instantly made off and disappeared. He tried this experiment several times, on the approach of whales, but always with the same result."† From this

---

* " Philosophical Transactions." vol. lxxviii. p. 414.
† " Histoire Naturélle des Cétacées" par M. Le Comte La Cépède.

we may conclude that these animals have thus a perception, even at a great distance, of odoriferous bodies.

THE PECTORAL AND DORSAL FINS.

The fins are placed on each side, and contain bones similar to the anterior extremity of the digitated animals, inasmuch as there are bones corresponding to the arm, fore-arm, carpus or wrist, and a series of phalanges, which are enveloped within a strong condensed adipose membrane of a semi-cartilaginous consistence.*

Some of the other species (which I shall hereafter describe) have a dorsal fin, the forms of which greatly differ, and which is situated either at a greater or less distance from the orifice of the spiracles, or from the extremity of the tail. The common and Iceland whales have no such thing.

CONSTRUCTION OF THE TAIL.†

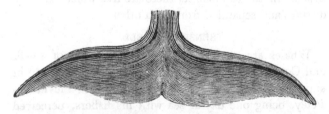

The manner in which this tremendous and only weapon of defence belonging to this animal is constructed is, perhaps, as beautiful as to its mechanism as any other part of the animal: it is wholly composed of three layers of tendinous fibres, covered by the com-

---

* The description of the skeleton of the Balænóptera Rórqual, or broad-nosed whale, which will be found in its proper place, will give the reader some idea of the formation of the skeletons of the Cetàcea generally.

† The above engraving represents a specimen which was twenty feet, five inches in breadth, in a whale fifty-three feet long.

mon cutis and cuticle : two of these layers are external, and the other internal. The direction of the fibres of the external layers is the same as in the tail, forming a stratum about one-third of an inch thick; but varying, in this respect, as the tail is thicker or thinner. The middle layer is composed entirely of tendinous fibres, passing directly across, between the two external ones above described, their length being in proportion to the thickness of the tail; a structure which gives amazing strength to this part.

The substance of the tail is so firm and compact, that the vessels retain their dilated state even when cut across; and this section consists of a large vessel surrounded by as many small ones as can come into contact with its external surface: which of these are arteries and which veins is not yet ascertained.

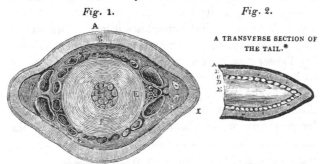

Fig. 1.　　　　　　　*Fig.* 2.

A TRANSVERSE SECTION OF
THE TAIL.*

The above cuts are representations of the posterior part of the animal, being a section of one of the lumbar vertebræ and a portion of the tail.

---

* EXPLANATION OF THE FIGURES.—FIG. 1.

A.—Skin with Epidermis.

B.—Cellular substance called blubber.

C.—Cartilage enveloping the tendinous cells.

D.—Cells of strong muscular coating, through which the tendons play.

GENERAL REMARKS ON THE ECONOMY OF THE WHALE.

All the species of whales do not content themselves with the species of food which I have already mentioned,* for some of the balænóptera prey upon fish of a tolerable size, and particularly such as assemble in troops, as mackarel, herrings, &c.†

Immediately beneath the skin lies the blubber, which encompasses the whole body of not only these, but all cetàceous animals. To speak *anatomically*, it is the adeps or fat, and lies exterior to the muscular flesh. In the porpoises, seals, trichecus or morse, and the narwhale, it is firm and full of fibres, and about an inch thick more or less. In the whale, its thickness varies, but ordinarily measures from eight to ten or twenty inches. The greatest quantity I ever saw cut from a whale was about forty tons; but sometimes it is so enormous as to produce fifty, eighty, or even one hundred tons; being lighter it swims on water. Its principal use appears to be partly to poise the body, and render it equiponderent to the water, partly to keep off the water at some distance from the blood, also to keep the whole warm by reflecting or reverberating the hot steams of the

---

E.—Spinal canal, enclosing a fascia of blood-vessels.

F.—Cartilaginous substance between the joints of the spine.

G. H.—Blood-vessels

I.—A ridge called the rump.

K.—Synovial glands.

Transverse Section of the Tail.—Fig. 2.

A.—Skin.

B.—Blubber.

C.—Tendinous envelope.

D.—Blood-vessels.

E.—Cartilaginous body.

Vide pages 23 and 24.   † Cuvier, vol. iv. p. 491.

body, and so redoubling the heat, as all fat bodies are less sensible to the impression of cold than lean ones.*

The blubber is well furnished with arterial blood, thus giving the fat somewhat of a pinkish appearance;† it is however in some animals a yellowish-white, yellow, or red. In the very young animals it is always a yellowish-white; in some old animals, it resembles in colour the substance of the salmon. The lips appear to be wholly composed of blubber, and will yield from one to two tons of pure oil each. The adeps on the tongue affords less oil than any other part of the body : in the centre of this organ, it is found mixed with muscular fibres. The lower jaw (with the exception of the jaw-bones) is almost entirely formed of pure fat; on the crown-bone there is a very considerable coating. The fins are principally blubber, bones, and tendons, and there is a very thin stratum of fat upon the tail.

The blubber is found on examination to be contained, like the fat of other animals, within the cells of the cellular or adipose membrane, which are connected together by a powerful reticulated combination of tendinous fibres. These fibres are condensed at the surface, and appear to form the substance of the skin. The oil is expelled when heated, and in a great measure discharges itself out of the cells whenever putrefaction in the fibrous parts of the blubber takes place. The blubber and the baleen, or whalebone, are the parts of the whale to which the attention of the whaler is particularly directed. The bones and flesh (excepting occasionally the inferior jaw-bones) are rejected. In its recent state, the blubber is perfectly free from any unpleasant smell;

---

* Dr. Rees's " Cyclopedia,"—Art. *Blubber.*
† Dewhurst's " Dictionary of Anatomy."

and it is not until after the termination of the voyage, when the cargo is unstowed, that a Greenland ship becomes disagreeable. Four tons of blubber, by weight, generally afford three tons of oil* when boiled, which is done in the port, on the arrival home of the vessel ; but the blubber of a sucker, or cub, contains a very small proportion. Whales have been caught that afforded nearly thirty tons of pure oil, when the blubber was boiled, and those yielding twenty tons of oil are by no means uncommon. The quantity of oil yielded by a whale generally bears a certain proportion to the length of its longest blade of baleen or whalebone. The average quantity is thus expressed by Captain Scoresby in the following table :

Length of Baleen in ft.	1	2	3	4	5	6	7	8	9	10	11	12
Oil yielded in gallons.	1½	2¼	2¾	3¼	4	5	6½	8½	11	13½	17	21

Although this statement averages very nearly the truth, yet sometimes exceptions occur. A whale of 2½ feet baleen has been known to produce nearly ten tons of oil, and another of twelve feet of baleen only nine tons. Such instances however are very uncommon.

A stout whale, of sixty feet in length, is of the enormous weight of about seventy tons; the blubber weighing about thirty tons; the bones of the head, baleen, fins, and tail, about eight or ten; and the carcase thirty or thirty-two.

The flesh of a young whale is of a red colour, and when cleared of fat, broiled and seasoned with salt and

---

* The ton or tun of oil is 252 gallons wine measure ; at a temperature of 60° Fah. it weighs 1933 lbs. 12 oz. 14 dr. avoirdupois.

pepper, eats not unlike tough beef; that of the old
whale approaches to black, and is exceedingly coarse.

Most of the bones forming the skeleton of the whale
are very porous, and contain large quantities of fine oil.*
The inferior jaw-bones, which measure from twenty to
twenty-five feet in length, are often taken care of on
account of the oil that drains out of them, when they
come into a warm climate. When exhausted of oil they
readily swim in water. The external surface is the
most compact and hard. The ribs are pretty nearly
solid, and, according to the late Sir Charles Giesecké,
the ribs are thirteen on each side, but the skeleton of
the B. Rorqual, lately exhibiting at Charing Cross, had
fourteen.

From the peculiarity of structure in the fin, the Rev.
Dr. Fleming has named them *"swimming paws."*†

The posterior extremity of the whale as already mentioned is the tail, which is the termination of the spine
into the ossa coccygis, which runs through the middle of
it, almost to the edge.

Few opportunities of examining the internal structure
and peculiarities of the B. Mysticètus occur ; hence,
what is known respecting its anatomy is deduced principally from its analogy to other cetàceous animals of
the same genus, although of a smaller species, which
have come accidentally up the rivers, or entered the
harbours of Great Britain.

There is no certainty respecting the longevity of the
whale. It may be presumed, however, that individuals of the larger species may have lived (according to
the opinion of Baron Cuvier) more than a thousand

* The oil obtained from the interior of the bones is considered far superior to that derived from the blubber : it is, in fact, the " marrow."
† Fleming's " Philosophy of Zoology," &c

years; should this be any thing like accuracy, we need not be surprised at the genius of allegory adopting them as the emblems of duration.

I have already made a few observations respecting the length of this species of the Cetàcea :* the following table will however illustrate the dimensions of the various parts of the common whale.

A table of the Comparative Dimensions of six Mysticète, measured by Capt. Scoresby, Jun.

Portion of the Whale.	Ft.	In.	Ft.	In.	Ft.	In.	Ft.	In.	Ft.	In.	Ft.	In.
Longest blade of baleen	1	0	6	0	10	10	11	2	11	6	13	7
Extreme length	17	0	28	0	51	0	50	2	58	0	52	0
Length of the head	5	0	8	6	16	0	15	6	19	0	20	0
Breadth of under jaw ..	...	...	...	...	...	...	9	6	12	0	...	...
Length of tip of lip to fin	5	6	10	0	...	...	18	0	...	...	...	...
—— to greatest circumf.	7	0	...	...	...	...	24	0	...	...	...	...
Circumf. at the neck ..	10	0	18	6	...	...	31	6	...	...	34	0
Greatest circumference.	12	0	20	0	...	...	34	0	35	0	...	...
Circumf. by the genitalia	9	0	15	6	...	...	19	0	...	...	...	...
—— near the tail..	2	11	4	0	6	6	6	8	...	...	...	...
Fin — { Length......	2	3	...	...	7	0	6	4	8	6	9	0
Fin — { Breadth	1	3	...	...	4	0	4	0	5	0	...	...
Tail — { Length......	...	...	...	...	5	6	5	6	6	0	6	0
Tail — { Breadth	...	...	...	...	20	0	17	6	24	0	20	10
Lip — { Length......	4	9	8	2	15	6	15	0	18	6	19	6
Lip — { Breadth	...	...	...	...	...	...	...	...	...	...	6	2
Produce in oil (tons) ..	1		4		16		16		19		24	
Sex	Female.		Male.				Female.				Male.	

* *Vide* pp. 13, 14.

AN ICEBERG, IN LATITUDE 73° 32′ N.

METHOD EMPLOYED TO CAPTURE THE WHALE.

"The Almighty, having created man in *his own image,* gave him power and dominion over all the living creatures he had made, and to subdue them according to his several wants and necessities."— GEN. i. 26—28.

Having described the zoological peculiarities of the *balæna mysticètus,* it will perhaps not be deemed uninteresting to the reader if I give a concise detail of the manner in which these leviathans of the deep are captured, and thus rendered subservient to the wants of man.

The first object is of course to fit out a vessel suited to the trade, and able to encounter the vicissitudes to which she becomes exposed in these dreary and inhospitable regions. During the periods when the whale was captured in the bays, or on the margins of the icy fields, very slight vessels were sufficient; but as soon as the adventurous spirit of the whalers began to arrive in these seas at a very early period of the season, and make their way into the midst of immense tracts of floating ice and icebergs, it became important to protect the ships against the severe shocks and concussions to which their situation rendered them liable.

To effect this, a ship should be built in such a manner as to possess a greater degree of strength than ordinary. The exposed parts, which are generally at the head and bow of the vessel, should be secured with double and even with treble timbers; and likewise fortified externally with large iron plates, also internally with powerful stanchions and crow-bars, so disposed as to cause the pressure on any one part to bear upon, and be supported by the whole fabric. These vessels are not copperbottomed, as is the case with those engaged in the

southern or spermaceti whale fishery, and light vessels generally, inasmuch as the latter would not be sufficiently strong to resist the blows it receives from the
ice.

Captain Scoresby recommends for this trade the employment of a vessel of the dimensions of 350 tons register as the most eligible, but generally the tonnage
varies from 250 to 300. The Neptune of London, the
vessel in which I sailed, was 291 ; a ship of 350 tons is
occasionally filled (but this is an event that has rarely
happened for several years, and is always unlooked for),
the number of men required for its navigation, being
likewise necessary for manning the boats employed in
the fishery, and consequently could not well be reduced
in a much smaller vessel. A larger tonnage than the
one just mentioned is scarcely ever filled, and involves
the owners in useless extra expense. The Dutch merchants are of opinion that the vessels destined for this
employment should be at least a hundred and twelve feet
long, twenty-nine broad, and twelve deep, carrying
seven boats, and from forty to fifty seamen.

The whale ships destined for Greenland generally
leave their respective ports about the end of March or
commencement of April, but those intended for Davis's
Straits towards the end of February, or early in March,
and proceed either to the Shetland or Orkney Islands,
where they receive their complement of men, one-half
of which are engaged from one of these counties.* The
latest period of their arrival in the Polar Seas is the
latter end of April, and there being no equatorial line to
cross, as in the southern latitudes, the Greenland sea-

* The Orkney and Shetland Isles are the extreme northernmost counties of Scotland.

men perform a similar initiatory ceremony on this occasion, which perhaps may not prove unentertaining if I describe it, inasmuch as it was witnessed by myself in 1824, when I was duly baptized, and made one of sturdy Neptune's sons.* After having undergone this laughable custom, I was

" Duly hailed a free-born British Tar,
The SOVEREIGN of the SEA." †

Previous to the ship leaving her port, the seamen collected from their wives and other female friends a profusion of gaudy ribands for " the garland," of which great care was taken until a few days previous to the first of May, when all hands (officers excepted) were engaged in preparing the said garland, with a neat model of the ship.

The garland was made of a hoop taken from one of the beef casks, and fastened to a stock of wood of about four feet in length, and the model of the vessel, which had been made by the carpenter, so as to answer the purpose of a vane. It was then hoisted on a rope, between the main and mizen masts, where it remained the whole of the voyage.

The first of May having arrived, the nautical tyros (or those who had not been previously to the Arctic Seas) were kept from between the decks, and all intruders excluded, whilst the principal performers got ready the necessary apparatus and dresses.

The barber who officiated on this important occasion was an old weather-beaten Northumbrian seaman, one George Brown, the boatswain, and his mate, John Put, of Deptford, the cooper, who had fixed upon the grating

---

* It is difficult to say whence this custom arose, or the time when it was first practised ; in all probability it was co-eval with the commence- ment of the whale-fishery. † Old Song.

E

over the entrance of the fore-hatchway, the following inscription in chalk letters:

> NEPTUNE'S EASY SHAVING SHOP,
> KEPT BY
> JOHN JOHNSON,
> Boatsteerer.

The procession, being formed on the forecastle, moved onwards to the quarter-deck (where the captain and officers received them), in the following order:—

First,—the fidler, playing as well as he could on an old violin,—" *See the Conquering Hero Comes;*" next four men, two abreast, disguised with matting, rags, &c., so as to completely prevent their being recognized, each armed with a boat-hook, as a staff of office; then came the sovereign of the sea, Old Neptune, also disguised, ornamented with a paper crown, and mounted on the carriage of the largest gun in the ship, followed by the barber, barber's mate, 'shaving-box carrier, swab-bearer, and as many of the ship's company that chose to join them, dressed in such a grotesque style as completely to put it out of the power of language to describe. Arrived on the quarter-deck, they were met by the captain and officers, when his marine majesty condescended to dismount, and the following dialogue ensued:—

Neptune.—Sir, are you the captain of the ship?

Captain.—I am.

Nept.—What's the name of your ship?

Capt.—The Neptune, of London.

Nept.—Where is she bound to?

Capt.—The Greenland Seas.

Nept.—What is your name?

Capt.—Matthew Ainsley.

Nept.—You are engaged in the whale-fishery?

Capt.—I am.

Nept.—Well, sir, I hope to have the gratification of drinking your honour's good health, and wish you a prosperous fishery.

Here the second mate, William Ford, presented the assembly with three cans, containing in the whole about three quarts of rum.

Nept.—(Filling a glass) Captain, here's a health to you, and success to our cause. Have you got any freshwater sailors on board? if you have, I must christen them, in accordance with our laws, so as to make them useful to ourself and country.

Capt.—There are eight on board who have made this their first voyage; they are at your service: I therefore wish you good morning.

The procession then returned in the same manner as it came, with the addition of the candidates for nautical fame, who followed in the rear: after descending the fore-hatchway they congregated between decks; Neptune's operations being performed at the foot of the ladder; here the offerings to Neptune were given to his receiver-general, *i. e.* the cook, and which consisted of whiskey, rum, tobacco, tea, &c. The barber then stood ready with his box of lather, and the landsmen being ordered before Neptune, the following dialogue took place, only with the alteration of the man's name :—

Nept.—What is your name?

Ans.—Gilbert Nicholson.

Nept.—Where do you come from?

Ans.—Lerwick.

Nept.—Where is that?

Ans.—In Shetland.

Nept.—Have you ever been to sea before?

Ans.—No.

Nept.—Where are you going to ?

Ans.—Greenland.

Nept.—For what purpose.

Ans.—To help you to catch whales.

At each of these answers, the brush being well dipped by the barber in the lather (which consisted of a mixture of oil, tar, paint, &c.), was thrust into the respondent's mouth and over his face; then the barber's mate scraped his face with a *razor, i. e.* a piece of iron hoop in the form of that instrument, well notched; his lacerated face was wiped with a *damask towel* (a boat-swab, dipped in the most filthy water), and this ended the ceremony with each individual. When completed, the disguises were removed, the fidler struck up some lively tunes, the grog was pushed plentifully about, and dancing continued until most of the crew *were full three sheets in the wind,* or, in plain language, nearly intoxicated.*

After this digression, I may observe, that formerly it was the custom of the whalers to stay a few weeks at what is denominated the " Seal-fisher's Bight," extending all along the Greenland coast, previous to their pushing in those more northern waters, where, amidst large fields or floes and mountains of ice, the powerful and precious mysticètus is tossing; but in later times it has become usual to sail at once into that centre of danger and enterprise.

Long before the seamen have arrived in the country, or " on fishing ground," all hands are daily at work in making the necessary preparations, so that at the proper time no delay may take place.

---

* This account of the " *May Day Christening,*" formed the subject of an article I inserted in Hone's Table Book for 1827, a volume abounding with much curious and valuable information on local customs and events.

Previous to the time of the
elder Captain Scoresby, captains or harpooners on watch,
at the mast head, were only
protected from the inclemency
of the weather by a bit of canvass; but, this being found extremely inconvenient, this gentleman constructed what is
technically called "*the Crow's
Nest*," which is as simple as
ingenious, consisting merely of
a species of sentry-box, made of
light wood in the shape of a
cask, having a seat in the middle, and a species of trap-door
in the floor; this is provided
with a telescope, a speaking-
trumpet, and a signal instrument of this shape, denominated

CROW'S NEST.

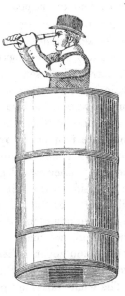

THE YONDER.

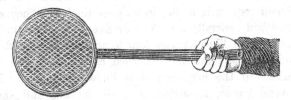

This signal instrument (the yonder), is simply a hoop,
with canvass stretched across, and attached to a handle
about four feet long; sometimes a fowling-piece, with a
bag of shot and a horn of powder. Thus furnished, it
is placed on the main-top-mast, and main-top-gallant-
mast-head, many vessels having them on the fore-top-
gallant-mast-head. This is the post of honour, and I

need hardly say of severe cold too, and the place where the captain or harpooners frequently sit for hours together, in a temperature thirty or forty degrees below the freezing point. From this place an immense extent of country can be descried, with all the movements of the surrounding sea and ice, as also the appearance of whales or any other animals peculiar to this climate, none of whom could be perceived from the deck.

As soon as the whalers arrive in the Polar Seas, which are frequented by the object of their adventure, the crew is constantly on the alert, keeping watch night and day. Each vessel generally has seven boats, three on each side of the ship, suspended from davits, and the jolly-boat over the stern; so that if required they can be lowered at a moment's warning : and in calm weather, when there are many whales near, one is usually kept manned and afloat, either attached to the stern or rowing a small distance from the ship.

The whale-boats are painted at both ends; the names of the ship and the captain, with the date of the year, are painted on the bows; and the keel is protected with iron, in case of accident or other circumstances that should require it to be drawn over the ice. They are furnished generally with four or five oars, and one for the steerer; likewise from four to six lines of a hundred and forty fathoms of rope, each coiled down and attached to the harpoon; it is likewise supplied with several lances, a hatchet, a small red flag and staff, sometimes a fowling-piece with powder-horn and shot-bag; the latter, however, are never used for the purpose of capturing the whale; but in order to attack or defend themselves from the Polar bear, or shoot any of the numerous birds inhabiting this climate, which forms a most nutritious and delicate species of food.

INSTRUMENTS USED IN THE CAPTURE AND DISSECTION
OF THE WHALE.*

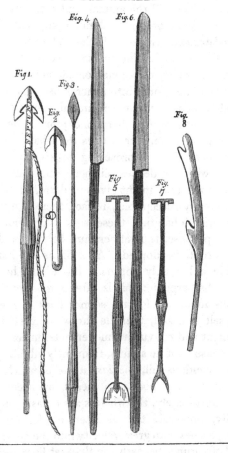

* EXPLANATION.

Fig. 1. The hand-harpoon.
— 2. The gun-harpoon.
— 3. The lance.
— 4. The tail-knife.
— 5. The blubber-spade.

Fig. 6. The blubber-knife.
— 7. The pricker.
— 8. The mik´ or rest for the harpoon, generally placed at
the head of the boat.

The harpoon is the principal instrument used to catch the whale, and is employed to fix the animal to the line; not to pierce and kill the animal, but merely to prevent its escape. Some of the whalers employ what is called the *gun-harpoon*, which is discharged from a peculiar species of gun; but owing to the noise of the report in its discharge, which occasions considerable alarm to the whale, it has not come into general use : consequently the hand-harpoon is the one commonly employed, and is generally thrown by the harpooner; or, if he is sufficiently near, he at once transfixes the whale with it.*

When boats are approaching a whale, they row towards it with the deepest silence, cautiously avoiding the least noise that might give an alarm, of which the whale is greatly susceptible; in fact, there is not a more timid animal inhabiting these seas. Occasionally, to effect this purpose, they attack him behind, and a very circuitous route is adopted. As soon as the harpooner can safely and quietly approach the whale, he attacks him with the harpoon : this is often a most critical and dangerous moment; for no sooner does this leviathan feel himself attacked, than he throws himself into the most violent and convulsive movements, in consequence of the intense pain he suffers, frequently vibrating in the air his tremendous tail, *one dash alone* of which is sufficient to founder a boat into pieces, an occurrence to which, unfortunately, the whalers are occasionally liable. Generally, however, the monster, on feeling himself wounded, dashes with great velocity into the depths of the sea, and oftentimes beneath the thickest floes and moun-

---

* The Congreve and other species of rockets have been unsuccessfully employed to destroy the whale, as also an improved harpoon invented by Captain Manby, F. R. S. But these are rarely employed, except experimentally.

tains of ice : when in this situation, he moves frequently at the rate of *eight or ten miles an hour*; therefore, it obliges the harpooner to employ the utmost diligence, in order that the line attached to the harpoon may run out smoothly and without the least entanglement; for, should this unfortunately occur, the whole crew and the boat would be liable to be dragged after him under the waves. The boat that has become thus fast to the whale is commonly followed by a second, for the purpose of not only furnishing assistance, if required, but likewise to supply more line, as the first is generally run out in the course of eight or ten minutes. Should, however, the second boat be at a distance at the time her aid is required, the crew of the first hold up one, two, or three oars, in order to intimate their pressing need of supply; at the same time turning the rope once or twice round a species of post, denominated the *bollard* or *billet-head*, at the head of the boat, by means of which the motion of the line and the career of the animal are somewhat retarded; this, however, becomes a very delicate operation; for sometimes the head or side of the boat is actually brought to the very edge of the water; and if the rope is drawn at all too tight, and prompt assistance is not afforded, it will sink it altogether.

During the time the line is whirling round the bollard, the harpooner is compelled to keep throwing water over it; for the friction is so great, that sometimes, notwithstanding this precaution to prevent its catching fire, he frequently becomes enveloped with smoke. If, after all, they are left without assistance, they let out the whole line; thereby not only losing their prize, but likewise the harpoon and all the lines.

The length of time between the period of the first striking the whale and effecting her death, averages from

three to four hours. During the time I was in Greenland, we caught five whales, and I shall here insert a few
extracts respecting them from my journal.

" *Latitude by Observation,*
80° 31′ *N.*

" 1824. MONDAY, MAY 10. *Wind, E. N. E.*—The
first and middle part, moderate breezes and clear weather; the ship dodging amongst great quantities of small
loose ice, numbers of straggling whales to be seen; two
boats constantly on the watch. About six P. M. Mr.
John Larnders (chief mate) struck our first whale; at
seven P. M. she was killed; and at ten towed alongside:
at eleven the crew began flensing her, which was finished
at four A. M. of the 11th. The jaw-bones were taken
in, and the baleen (whalebone) measured six feet seven
inches long."

" *No Observation taken.*
" THURSDAY, MAY 13. *Wind, N. N. W.*— Fresh
breezes, at times snow; four whales seen from the mast-
head, all hands called up; lowered away four boats,
when shortly afterwards William Ford (second mate)
struck a second whale, at one A. M., which, after running
out seven lines (about 900 fathoms, or 5,400 feet), the
harpoon drew out, and we lost her.

" At nine A. M. all hands again called; and at ten
minutes to one P. M. John James, the loose harpooner,
struck a third whale; at a quarter past two P. M. she
was killed, and towed alongside at three : the flensing
began at five o'clock, and was finished by seven.

" At ten P. M., the wind being N.N.E., William Ford
struck a fourth whale; she was killed at eleven, and
towed alongside at twelve (midnight), when the flensing
immediately began."

" *No Observation taken.*
" FRIDAY, MAY 14. *Wind, N.N.E.*—At two o'clock

A. M. they finished flensing the last whale; the morning was most remarkably cold, with heavy showers of snow, fresh breezes at the first part of the day; the decks were cleaned, and the watch set. At four A. M. a whale was seen running to windward; two boats lowered away after her, when accidentally John James's boat capsized on a piece of ice; he hoisted a signal of distress, when two boats were sent off to his assistance, and he was brought off without any injury having been sustained either by the boat or crew. At six P. M. saw one whale, and lowered away two boats after her; at eight P. M. boats returned without success. The whalebone of the last two whales measured as follows: *viz.* the first, four feet two inches; the second, five feet four inches."

" *No Observation taken.*

" SATURDAY, MAY 15. *North.*—At two A. M. Mr. John Larnders struck the fifth whale; at three she was killed, and in half an hour she was towed alongside; at ten minutes past four they began flensing her, and at five A. M. they finished. At two P. M. saw one whale, but the wind blowing very strong from the North, consequently prevented any boats being lowered after her; this last whale was one foot less than the third. During the latter part very heavy gales set in, so that it was impossible to lower the boats; wind shifting to the N.N.E., we were compelled to ply under close-reefed topsails among heavy loose ice. The gales continued extremely strong for the next twenty-four hours; numerous whales in sight."

" *Latitude by Observation,*
78° 52′ *N.*

" MONDAY, MAY 24. *Wind,* N. N. E.—The first part fresh breezes, with very foggy weather; at six A. M. John James struck the sixth whale, which, after giving the crew a tiresome chase over ice as well as water, was killed at eleven o'clock, and in an hour after was towed alongside. The ship was made fast to a large floe of ice, by means of small ice anchors, similar in shape to the letter S : at two in the afternoon the flensing began, which was finished at seven. This whale took out fourteen lines of 140 fathoms each, consequently about 1,960 feet of heavy rope. During the time the boats were after her, William Ford spoke the Harmony of Whitby, whose Captain informed him of the loss of the Aimwell of Whitby, commanded by Captain Johnson, from her trying to get to leeward of a large piece of ice: this ship's side was stove in by a large pointed tongue (as the seamen quaintly call it) of ice, being unseen from the surface of the water ; the crew knew nothing of it, until the water had got up to the second tier of casks in the hold. The Esk of Whitby rescued three boats and their crews ; the Lively of the same port received also a boat and her crew ; the captain and his officers, by the Abram of Hull, Captain Jackson : fortunately no lives were lost. Beautiful clear weather the latter part of this day."

"*Latitude by Observation,*
70° 14′ *N.*

"WEDNESDAY, JULY 7. *Wind*, W. N. W.—The first and middle part, moderate breezes, at times extremely foggy ; running to the S. W. among very heavy ice. At meridian a whale was observed, and two boats sent away ; at half-past one John James made fast a harpoon, and she run out thirteen lines (about 1820 feet): at five P. M. she was killed in a hole formed in the ice. During the latter part, the weather became very foggy, and the people endeavoured to get the whale and lines heaved in by the capstan ; but, after many hours of useless labour, this could not be accomplished, and the whale was consequently lost. An immense number of finners (the *balænóptera jubartes*, La Cépède), and nar-whales were seen swimming about. The ship anchored to a vast floe of ice."

I have here given a brief detail of the periods occupied in the principal events connected with the capture of the whale which I myself witnessed; but there are a few circumstances connected with it, which it is my duty to notice.

The moment a whale is struck by the harpooner, and she makes her exit, one of the sternmost immediately hoists a small red flag, which is immediately answered by the union-jack being hoisted on board the ship at the mizen-top-mast. This is likewise a signal to other vessels that the ship's boats are fast to a whale, and that although, as is frequently done, they may assist in the capture, they have no right whatever to the whale. However, sometimes disputes on this subject occasionally occur, particularly if the whale rises near another ship's boats, and she is harpooned without the appearance of the first harpoon. Besides, when several ships are in company, and one or more whales are seen, all the ships immediately send away boats after her, and the boat whose flag announces her harpooner being fast, claims her as his prize. These disputes are sometimes obliged to be settled by referring to a court of law, as to which ship should possess the property or its value.

The signal flag of the boat being observed by the ship, the whole crew are awakened by the watch on deck crying loudly, " A fall! a fall!"* On this being heard, the seamen do not allow themselves even time to dress, but rush out in their shirts and drawers into an atmosphere

---

* Derived from the Dutch word *val*, expressive of the precipitate haste with which the sailors get into the boats.

which is frequently below *zero*, carrying along with
them their clothes made into a bundle, with a piece of
rope-yarn, trusting to find an opportunity when in
the boats of dressing themselves, or during the time
the boats are lowering. There is such a degree of tu-
mult and noise on this occasion, that Captain Scoresby
informs us he has known young mariners raise cries of
alarm, thinking the ship was going down : and a recent
writer states, in one instance the panic was so great, that
death speedily followed.* The alarm of " a fall " has a
singular effect on the feelings of a sleeping person, unaccustomed to the whale-fishing business. It has often
been mistaken as a cry of distress. A landsman, in a
Hull ship, seeing the crew, on an occasion of a fall,
rush upon deck, with their clothes in their hands, and
leap into the boats, when there was no appearance of
danger, thought the men were all mad.

The period that a whale remains beneath the surface
of the water varies; but Captain Scoresby estimates it
at half an hour. Pressed for respiration, he re-appears
above, generally at a considerable distance from the spot
where he was first harpooned, in a state of great ex-
haustion, which is ascribed to the severe pressure en-
dured when placed beneath a column of water of 700 or
800 fathoms deep, and which must be equal to 211,200
tons. When a boat is fast to a whale, the others row
about in various directions, that one at least may get a
start, as it is denominated, about 200 yards off the point
of his rising, at which distance they can easily reach and
pierce him with one or two more harpoons prior to his
again descending, which is usually the case for a few

---

* " Narrative of Discovery and Adventures in the Polar Seas and
Regions," 1831. Page 384.

minutes. When he re-appears, he is then fiercely attacked with lances, by means of which he becomes wounded in the most vital parts. Blood, freely mixed with oil, streams copiously from his wounds and the blow-holes, dying the sea for a considerable distance round, sprinkling and sometimes copiously drenching the boats and their crews. By this time the animal becomes more and more exhausted : but oftentimes, at the approach of death, he makes a powerful, convulsive, and energetic struggle, rearing his tail high and aloft in the air, whirling it with a noise which is heard at the distance of several miles. At length, perfectly overpowered and exhausted, he lays himself on his back or his side, and then expires. The flag is then taken down, and three loud huzzas raised by all the boats' crews. No time is lost in piercing the tail with two holes, through which ropes are passed, which, being fastened to the boats, tow the animal to the vessel, amidst shouts of merriment and joy.

As soon as the whale is caught and secured to the sides of the ship, another important operation is performed, which is that of flensing, or extracting the blubber and whalebone. Now, if the strength of the crew is put into requisition, this can be performed in about four hours, although a much longer time is frequently occupied. Previous to their commencing, the captain or surgeon stands at the companion-door, and calls every seaman to him, giving each a dram of rum, with a double allowance to the important personages denominated the kings of the blubber, whose office it is to receive that precious commodity and stow it carefully in the hold. Another great officer is called the *specksioneer*, and to him is confided the direction and management of all the cutting operations. The first step is to form round the

whale, between the neck and fins, a circle denominated
the *Kent,* around which all proceedings are to be conducted: to it is fastened a machinery, composed of
blocks and ropes, called the *Kent-purchase ;* by which,
with the aid of a windlass or capstan, the whale can be
turned on all sides, and in any required direction. The
harpooners then, directed by the specksioneer, commence with spades and powerful knives (these last are
frequently made from old sword and cutlass blades) to
make long parallel cuts from end to end, which are divided into cross pieces of about half a ton: these are
hoisted upon deck by tackle attached to the *blubber-
guy,** where others again reduce them into still smaller
pieces.   Finally, these are thrown down the main hatch-
way, into the *flens-gut,* where, being conveniently stowed
away by the *kings,* they remain until time can be spared
for the whole (in a day or two after) to be again thrown
upon deck, when it is again subdivided into pieces of
about three inches wide, and somewhat about six long,
and with an instrument, called a *pricker,* it is packed
carefully through the bung-hole into the casks.

As soon as the cutting officers have cleared the whale's
surface of the carcass of the blubber that is above the
surface of the water, which does not exceed a fourth or
fifth part of the animal, the *Kent-machinery* is applied,
and turns the body round, until another part, yet untouched, is presented: this being also cleared, the whole
mass is again turned, and so on, until the whole has been
exposed, and the blubber removed.   The *Kent* itself is
then stripped, and the lower jaw-bones, with the baleen
or whalebone, being conveyed on board, there remains
only the *kreng,* nothing more than a huge shapeless

* A large piece of cable-rope fastened to the main-mast and fore-mast.

mass of bones, covered with a quantity of black muscular substance; which is abandoned, either to sink or
to be devoured by the flocks of ravènous birds and
sharks which duly attend on this occasion.

However, when the fishery was transferred to the icy
banks in the open sea, this operation was necessarily
deferred till the cargoes were deposited in the Dutch or
British ports.

In the early days of the whale-fishery, when the whales
were found in great numbers immediately around the
shores of Spitzbergen, the Dutch formed a settlement on
that island, and performed there all the operations of
preparing the bone and extracting the oil from the
blubber. To so flourishing an extent was the fishery at
this time (the latter part of the seventeenth century)
as carried on by that nation, that they actually erected a
village on this desolate coast, all the houses of which
were brought, ready prepared, from Holland. They gave
it the name of *Smeerenberg* (from *smeeren,* to melt).
" This," says Mr. Macculloch, " was the grand rendezvous of the Dutch whale-ships, and was amply provided
with boilers, tanks, and every sort of apparatus required
for preparing the oil and the bone. But this was not all:
the whale fleets were attended by a number of provision-
ships, the cargoes of which were landed at Smeerenberg,
which abounded during the busy season with well-
furnished ships, good inns, &c., so that many of the
conveniences and enjoyments of Amsterdam were found
within eleven degrees of the pole! It is particularly
mentioned, that the sailors and others were every
morning supplied with what a Dutchman regards as
a very great luxury—*hot rolls* for breakfast. Bata-
via and Smeerenberg were founded nearly at the
same period, and it was for a considerable time doubted

F

whether the latter was not the most important establishment.

EARLY HISTORY OF THE BRITISH WHALE-FISHERY.

From the narrative of the voyage of Ohthere the Dane, given by King Alfred, in his Saxon translation of Orosius, it would appear that the pursuit of the whale was practised by the people of Norway at least as early as the ninth century. Other northern authorities bear testimony to the same fact. Of the manner, however, in which the whale-fishery was carried on at this remote era, we know nothing. It probably was not pursued on any systematic plan, but merely in the way of occasional encounters, as the hunting of wild animals on land would be practised in the same state of society. The inhabitants of the coasts surrounding the Bay of Biscay seem to have been the first who engaged in whale-fishing with a view to commercial purposes. They are, therefore, properly to be considered the originators of the pursuit as a branch of national enterprise. Their prosecution of it in the adjacent seas can be traced back as far as the twelfth century. The animal against which they directed their attacks, however, was most probably of a different species from that found in the Northern Ocean, and of a much smaller size. It seems to have been captured principally, if not exclusively, for the sake of its flesh, which was in those days esteemed as an article of food, the tongue especially being accounted a great delicacy.

By degrees, however, the number of whales that resorted to the Bay of Biscay diminished, and at length the whales altogether ceased to visit that sea. In these circumstances the Biscayan mariners carried the navigation farther and farther from their own shores, till at last they approached the coasts of Iceland, Greenland, and Newfoundland. Thus was commenced, in the course of the sixteenth century, the northern whale-fishery, as pursued in modern times.

The earliest whaling voyage made by the English appears to have taken place in the year 1594. The merchants of Hull are recorded to have fitted out ships for the fishery in 1598; and much about the same time the Dutch engaged in the trade. The Hamburghers, the French, and the Danes quickly followed At first, both in England and Holland the business was carried on by companies which had obtained charters for its exclusive prosecution. At length, however, it was thrown open in both countries to individual enterprise, under which new system it was found to be conducted with much more success and profit. The Dutch monopoly was put an end to in 1642; the English not till long afterwards. In this country, indeed, the trade was in the hands of an exclusive company till about a century ago. Up to that date it had, in general, been attended only with loss to each successive association that engaged in it.

In 1732 the British Parliament first adopted the plan of attempting to encourage and establish the trade, by giving a bounty to every ship which should engage in it. The bounty was at first twenty shillings a ton; but it was raised in 1749 to double that rate, upon which, says a late writer, " a number of ships were fitted out, as much certainly in the intention of catching the bounty, as of catching fish." The bounty, which was afterwards reduced to thirty shillings, was again raised to its former amount, and subsequently reduced, first to thirty shillings, then to twenty-five, and after that to twenty, but was at last altogether withdrawn in 1824. The trade is, at present, therefore, carried on without any artificial support. The Americans, Hamburghers, and Prussians are now almost the only competitors with whom the English whalers have to contend. The French revolution, and the wars by which it was followed, drove both France and Holland from the field; and neither of these countries have succeeded in the attempts they have made since the peace, to re-enter upon a line of enterprise, their pursuit of which had been so long interrupted.

PROBABLE SUCCESS OF THE WHALE-FISHERY, IN THE ARCTIC SEAS.

The success of the fishery varies with the spot in which whales are found. The most advantageous that the Greenland Sea has afforded, has been considered to be on the border of those immense fields of ice with which a great extent of them is covered. In the open sea, when a whale is struck, and plunges beneath the waters, he may rise again in any part of a wide circuit, and at any distance from the boats; so that, before a second harpoon can be struck, he may plunge again, and by continued struggles effect his extrication: but, in descending beneath these immense fields, he is hemmed in by the icy floor above, and can only find an atmosphere to breathe by returning to their outer boundary. The space in which he can rise is thus contracted from a large circle to a semicircle, or even a smaller segment. Hence, a whale in this position is attacked with a much better chance of success : even two may be pursued at the same moment; a measure which, in the open sea, often involves the loss of both. In the flourishing state of the Dutch fishery, a hundred of their vessels have been seen at once, ranged on the margin of one of these immense

fields of ice, along which the boats formed so continuous
a line, that no whale could rise without being instantly
struck. At the same time it is to be remembered that
this situation was one of considerable danger, in consequence of the concussions and disruptions to which these
plains are liable.

When the ship is surrounded with floating fragments
of ice, the fishery, though difficult, is generally productive: but it is directly otherwise when those pieces
are packed together into a mass impervious to ships or
boats, yet leaving numerous holes, through which the
whale can mount and respire, without coming to the
open margin, or within the reach of his assailants.

The whalers, as soon as they perceive a whale blowing
through one of these apertures, alight upon the ice, and
run full speed to the spot, with a lance and a harpoon.
To attack the leviathan of the deep in this spot, and
under these circumstances, is perilous; and, even when
the whale is killed, the dragging his huge body either
over or under the ice to the ship is a most herculean
task, and frequently, in the last case, it cannot be
effected without cutting it into pieces.

When the great fields of ice, during the progress of
the season, are open at various points, the fishery becomes then liable to the same evils as occur among
packed ice. Still worse is the case when the sea is over-
spread with a thin newly-formed crust. The whale easily
finds or beats a hole through this covering, where neither
the boats can penetrate nor the men walk over it, without the most imminent danger. Yet Captain Scoresby
mentions a plan by which he continued to carry on his
movements, even over a very slender surface of bay-ice.
He tied together his whole crew, and made them thus walk
in a long line one behind another. There never fell in

more than four or five at a time, who were easily helped out by the rest. The sufferers had a dram to console them after their cold bath; and in fact this compensation was considered so ample, that master Jack was suspected of sometimes allowing himself to *accidentally* drop in with the view of being thus indemnified.

Another grand distinction respects, first, the Greenland fishery, which, generally speaking, is that already described, and is chiefly distinguished by the immense fields of ice which cover the ocean; and, secondly, the Davis's Straits' Fishery, where that element appears chiefly in the form of moving mountains, topping the deep. This last is arduous and dangerous, but usually productive : it commenced at a comparatively late period, since it is not mentioned by the Dutch authors prior to 1719; and Mr. Scoresby has been unable to ascertain the date when it was begun by the British. Within these few years it has experienced a remarkable extension; for now most vessels proceed thereto instead of Greenland, which is comparatively deserted.

AN ICEBERG IN LATITUDE 74° 34' N.

DANGERS ATTENDANT ON THE CAPTURE
OF THE WHALE.

THE dangers of the whale-fishery, in spite of the utmost care, and though under the direction even of the most experienced mariners, are imminent and manifold. The most obvious peril is that of the ship being beset and sometimes dashed to pieces by the approach and collision of those mighty fields and mountains of ice with which these seas are continually filled. The desolate and inclement region, which is the scene of enterprise, encompasses the pursuit with its worst hardships and dangers. In this realm of eternal winter, man finds the land, the sea, and the air, equally inhospitable. Every thing fights against him. The intensest cold benumbs his flesh and joints; while fogs or driving sleet often darken the sky, and at the same time arm the frost with a keener tooth. The ocean over which he moves, besides its ordinary perils, is crowded with new and strange horrors. Sometimes the ice lies extended in large fields that bar all navigation as effectually as would a wall of iron, and over whose rugged and broken surface he can only make his way but by leaping from point to point, at the risk of being ingulfed at every step. Sometimes it bears down upon him in vast floating fields with such an impetus that, at the shock, the strong timbers of his ship crack and give way like an eggshell, or are crushed and ground to fragments between two meeting masses. Sometimes it rises before him in the shape of a lofty mountain, which the least change in the relative weights of the portion above and that beneath the surface of the water may bring in sudden ruin upon his head, burying crew and vessel beneath the tumbling

chaos, or striking them far into the abyss. And as for
what may be dimly distinguished to be land, rimming
with its precipitous coasts these dreary waters, it may be
most fitly described in the lines in which the poet has
pictured one of the regions of the nether world.

> " Beyond this flood, a frozen continent
> Lies dark and wild, beat with perpetual storms
> Of whirlwind and dire hail, which on firm land
> Thaws not, but gathers heap, and ruin seems
> Of ancient pile ; or else deep snow and ice."

Almost the only vegetation that springs from this frost-
bound soil is a scanty verdure, formed of mosses, lichens,
and other low plants, that conceal themselves beneath
the snow. At the farthest limit to which adventure has
pierced, a night of four months' duration closes each
dismal year; throughout which human life has indeed
been sustained by individuals previously inured to a severe climate, but the horrors of which have, in most of
the instances in which the dreadful experiment has been
either voluntarily or involuntarily tried by the natives of
more temperate regions, only driven the wretched sufferers through a succession of the intensest bodily and
mental tortures, and then laid them at rest in the sleep
of death.

The Dutch writers mention many dreadful shipwrecks, of which I shall relate a few, from a contemporary author.*

Didier Albert Raven, in 1639, when on the border of
the Spitzbergen ice, was assailed by a furious tempest.
Though the ship was violently agitated, he succeeded in
steering her clear of the great bank, and thought himself
in comparative safety, when there appeared before him

---

* " Edin. Cabinet Library."   Vol. on the Polar Regions.

two immense icebergs, upon which the wind was violently driving his vessel. He endeavoured, by spreading
all his sails, to penetrate between them; but in this
attempt the ship was borne against one with so terrible
a shock, that it was soon found to be sinking. By cutting away the masts, the mariners enabled her to proceed; yet, as she continued to take in water, several
boats were launched, which, being overcrowded, sunk,
and all on board perished. Those left in the ship found
their condition more and more desperate. The fore-part
of the vessel being deep in the water, and the keel rising
almost perpendicular, made it extremely difficult to avoid
falling in the sea; while a mast, to which a number had
clung, broke, plunged down, and involved them in the
fate of their unfortunate companions. At length the
stern separated from the rest of the vessel, carrying with
it several more of the sailors. The survivors still clung
to the wretched fragments, but one after another was
washed off by the fury of the waves, while some, half
dead with cold, and unable to retain their grasp of the
ropes and anchors, dropped in. The crew of eighty-six
was thus reduced to twenty-nine, when the ship suddenly changed its position, and assumed one in which
they could more easily keep their footing on board. The
sea then calmed, and, during the respite thus afforded,
they felt an irresistible propensity to sleep; but to some
of them it was the fatal sleep of extreme cold, from
which they never awoke. One man suggested the construction of a raft, which was framed according to the
captain's advice; happily, no sooner was it launched
than the waves swallowed it up. The remnant of the
crew encountered the next night another severe gale;
and the sufferings of the crew, from cold, hunger, and
burning thirst, were so extreme, that death in every

form seemed now to be inevitable. Fortunately, in the morning a sail was discovered: their signals were understood; and, being taken upon board, twenty survivors, after forty-eight hours of this extreme distress, were ultimately restored to safety.

In 1760, the Blecker (Bleacher), Captain Pitt, was driven against the ice with such violence, that in an instant all her rigging was dashed to pieces. Soon after, twenty-nine of the crew quitted the vessel, and, leaping by the help of poles and perches from one fragment of ice to another, contrived to reach the main field. The captain, with seven men, remained on board, and endeavoured to open a passage; but soon after, the ship again struck, when they were obliged to go into a boat, and commit themselves to chance, the snow falling so thick that they could scarcely see each other. As the weather cleared, they discovered their companions on the ice, who threw a whale-line, and dragged them to the same spot. There, the party, having waited twelve hours in hopes of relief, at length trusted themselves to the boats, and in twelve hours more were taken up by a Dutch vessel.

Captain Bille, in 1675, lost a ship richly laden, which went down suddenly; after which, the crew wandered in boats over the sea for fourteen days before they were taken up : thirteen other vessels perished that year in the Spitzbergen seas. Three seasons after, Captain Bille lost a second ship by the violent concussion of the ice, the crew having just time to save themselves on a frozen field. At the moment of their disaster they were moored to a large floe, along with another brig, called the Red Fox; which last shortly after met a similar fate, being struck with such violence that the whole, hull and masts together, disappeared almost in an instant, the sailors of Captain Bille's company having had merely time to leap on the ice. The united crews now adopted various plans; some keeping their stations, others setting out in boats in different directions; but all, in one way or other, reached home. The same year the Concord went down in an equally sudden manner; but the crew were happily taken up by a neighbouring ship.

The whale-fishery is not more distinguished for examples of sudden peril and bereftment than for unexpected deliverance from the most alarming situations.[*]

The Davis's Straits' fishery has also been marked with very frequent and fatal shipwrecks. In 1814, the Royalist, Captain Edmonds, perished with all her crew; and in 1817 the London, Captain Matthews, shared the same fate. The only account of either of these ships ever received was from Captain Bennett, of the Venerable of Hull, who, on the 15th April, saw the London in a tremendous storm, lying to windward of an extensive chain of icebergs, among which it is probable she was dashed to pieces that very evening. Large contributions were raised at Hull for the widows and orphans of seamen who had suffered on these melancholy occasions.

Among accidents on a smaller scale, one of the most frequent is, that of the boats employed in pursuit of the whale being overtaken by deep fogs or storms of snow, which separate them from the ship, and never allow them to regain it. A fatal instance of this nature occurred to the Ipswich, Captain Gordon, four of whose boats, after a whale had been caught, and even brought to the ship's side, were employed on a piece of ice haul-

---

[*] Several interesting anecdotes are inserted in Professor Jamieson's, Leslie's, and Mr. Murray's works, lately published, to which the reader is referred.

ing in the line, when a storm suddenly arose, and caused
the vessel to drift away, and prevented her, notwithstanding the utmost efforts, from ever coming within
reach of the unfortunate crews who composed the greater
part of the ship's company. Several casualties of this
nature are related by Captain Scoresby, which occurred
to his boat's companies, all of whom, however, in the
end, happily found their way back to the vessel. One
of the most alarming cases was that of fourteen men
who were left upon a small piece of floating ice, with a
boat wholly unable to withstand the surrounding tempest; but, amid their utmost despair, they fell in with
the Lively of Whitby, and were most cordially received
on board.

The source, however, of the most constant alarm to
the whaler is connected with the movements of that
powerful animal, against which, with most unequal
strength, he ventures to contend. Generally, indeed,
the whale, notwithstanding his immense strength, is
gentle and even passive, seeking, even when he is most
hotly pursued, to escape from his assailants, by plunging into the lowest depth of the ocean. Sometimes,
however, he exerts his utmost force in violent and convulsive struggles; and every thing with which, when
thus enraged, he comes into collision, is dissipated or
destroyed in an instant. The Dutch writers mention
Jacques Vienkes of the Gört Moolen (the Barley Mill),
who, after a whale had been struck, was hastening with
a second boat to the support of the first. The whale,
however, rose, and with its head struck the boat so
furiously that it was shivered into pieces, and Vienkes
was thrown with its fragments on the back of the huge
animal. Even then, the bold mariner darted a second
harpoon into the body of his victim; but unfortunately

he got entangled in the line and could not possibly extricate himself, while the other party were unable to approach near enough to save him. At last, however, the harpoon was disengaged, and he swam to the boat.

In some instances, the boat, instead of being struck into the water, has met with the equally alarming fate of being projected by a stroke of the powerful animal's head or tail into the air. The following remarkable instance of this is given by Captain Scoresby, who, in one of his earliest voyages, saw a boat thrown several yards in the air, from which it fell on its side, plunging the crew into the sea. Happily, they were rescued from a watery grave, and only one was found to have received a severe concussion. " Captain Lyons, of the Raith of Leith," says our author, " while prosecuting the whale-fishery on the Labrador coast, in the season of 1802, discovered a large whale at a short distance from the ship. Four boats were despatched in pursuit, and two of them succeeded in approaching it so closely together that two harpoons were struck at the same moment. The whale descended a few fathoms in the direction of another of the boats, which was on the advance, rose accidentally beneath it, struck it with its head, and threw the boat, men, and apparatus about fifteen feet into the air. It was inverted by the stroke, and fell into the water with its keel upwards. All the people were picked up alive by the fourth boat, which was just at hand, excepting one man, who, having got entangled in the boat, fell beneath it, and was unfortunately drowned. The whale was afterwards killed."

In 1807, the crew of Mr. Scoresby, sen., had struck a whale, which soon re-appeared, but in such a state of violent agitation, that no one was able to approach it. The captain courageously undertook to encounter it in a boat by himself, and succeeded in striking a second harpoon; but, another boat having advanced too close, the animal brandished most furiously her tail with so much fury that the harpooner, who was directly under, judged it prudent to leap at once into the sea. The tail then struck the very place that he had left, and cut the boat entirely asunder, with the exception of two planks, which were saved by having a coil of ropes laid over them; so that, had he remained, he must have inevitably been dashed to pieces. Happily all the other seamen escaped injury. However, the results of these accidents are not always so fortunate. The late Aimwell of Whitby, in 1810, lost three men out of seven; and in 1812 the Henrietta, of the same port, lost four out of six, by the upsetting of the boats, the crews being thrown into the sea.

In 1809, one of the men belonging to the Resolution, of Whitby, struck a sucking whale; after which the mother, being seen wheeling rapidly round the spot, was eagerly watched. Mr. Scoresby, sen., being on this occasion in the capacity of harpooner in another boat, was selecting a situation for the probable re-appearance of the parent whale, when suddenly an invisible blow stove in fifteen feet of the bottom of his boat, which filled with water, and instantly sunk. The crew were saved.

Another and frequent misfortune is entanglement of the line, during the period the whale is retreating, which is often productive of disastrous consequences. A sailor belonging to the John of Greenock, in 1818, having happened to step into the centre of a coil of running rope, had a foot entirely carried off, and was obliged to have the lower part of the leg amputated. A harpooner belonging to the Henrietta of Whitby, when engaged in lancing a whale into which he had previously struck a harpoon, incautiously cast a little line under his feet, that he had just hauled into his boat, after it had been drawn out by the whale: a painful stroke of his lance induced the animal to dart suddenly downward, his line began to run out from beneath his feet, and in an instant caught him by a turn round the body. He had but just time to cry out, " clear away the line,"—" Oh dear!" when he was almost cut asunder, dragged overboard, and never seen afterwards. The line was cut directly, but without avail.

A whale sometimes produces danger by proving to be alive after having exhibited every symptom of death. Mr. Scoresby mentions the instance of one which appeared so decidedly dead that he himself had leaped on the tail, and was busy putting a rope through it, when he suddenly felt the animal sinking from beneath him. He made a spring towards a boat that was some yards distant, and, grasping the gunwale, was assisted on board. The whale then moved forward, reared his tail aloft, and shook it with tremendous violence, that it resounded to the distance of several miles. After two or three minutes of this violent exertion, he rolled on his side, and expired.

Even after life is extinct, all danger is not over. In the operation of flensing, the harpooners sometimes fall into the whale's mouth, with the imminent danger of being drowned. In the case of a heavy swell of the sea, they are drenched, and sometimes washed over by the surge. Occasionally they have their ropes broken, and are wounded by each other's knives. Mr. Scoresby mentions a harpooner who, after the flensing was completed, happened to have his foot attached by a hook to the kreng or carcase, when the latter was inadvertently cut away. The man caught hold of the gunwale of the boat;

but the whole immense mass was suspended by his body, occasioning the most excruciating torture, and even exposing him to the danger of being torn asunder, when his companions contrived to hook the kreng with a grapnel, and bring it to the surface.

It not unfrequently happens that after the exertions of many hours the whale makes its escape and is lost. Captain S. relates an extraordinary case of a whale struck on the 25th of June, 1812, by one of the harpooners belonging to the Resolution of Whitby, then under his command, which after a long chase broke off, and took with it a boat and twenty-eight lines, the united length of which was 6,720 yards, or upwards of three English miles and three-quarters. The value of the property thus lost was above one hundred and fifty pounds sterling; and the weight of the lines above thirty-five hundred weight. They soon after, however, again got sight of the animal near two miles off, and immediately re-engaged in the pursuit. They came up with it by great exertions about nine miles from the place where it was first struck. The attack was now renewed. " One of the harpooners," continues Captain Scoresby, " made a blunder; the whale saw the boat, took the alarm, and again fled. I now supposed it would be seen no more; nevertheless we chased nearly a mile in the direction I imagined it had taken, and placed the boats, to the best of my judgment, in the most advantageous situations. In this case we were extremely fortunate. The whale rose near one of the boats, and was immediately harpooned. In a few minutes two more harpoons entered its back, and lances were plied against it with vigour and success. Exhausted by its amazing exertions to escape, it yielded itself at length to its fate, received the piercing wounds of the lances without re-

sistance, and finally died without a struggle. Thus terminated with success an attack upon a whale which
exhibited the most uncommon determination to escape
from its pursuers, seconded by the most amazing
strength, of any individual whose capture I ever witnessed. After all, it may seem surprising that it was
not a particularly large individual, the largest lamina of
whalebone only measuring nine feet six inches, while
those affording twelve feet *bone* are not uncommon.
The quantity of line withdrawn from the different boats
engaged in the capture was singularly great. It amounted
altogether to 10,440 yards, or nearly six English miles.
Of these, thirteen new lines were lost, together with the
sunken boat, the harpoon connecting them to the whale
having dropped out before it was killed." There had
been eight boats in all engaged in this extraordinary
chase.

Of the dangers sometimes occasioned by the resistance
of the whale, or its efforts to retaliate upon its assailants, Captain Scoresby relates various instances. It
has happened that the harpooner has been struck dead
in an instant by a blow from the animal's tail. At other
times the stroke has fallen upon the boat and jerked the
crew out of it into the water. "A large whale," says
Capt. S., " harpooned from a boat belonging to the
same ship (the Resolution of Whitby) became the subject of a general chase on the 23d of June, 1809. Being
myself in the first boat which approached the fish, I
struck my harpoon at arm's length, by which we fortunately evaded a blow that appeared to be aimed at the
boat. Another boat then advanced, and another harpoon was struck; but not with the same result, for the
stroke was immediately returned by a tremendous blow
from the fish's tail. The boat was sunk by the shock,

and at the same time whirled round with such velocity that the boat-steerer was precipitated into the water, on the side next to the fish, and was accidentally carried down to a considerable depth by its tail. After a minute or so he arose to the surface of the water, and was taken up along with his companions into my boat. A similar attack was made on the next boat which came up; but the harpooner, being warned of the prior conduct of the fish, used such precaution that the blow, though equal in strength, took effect only in an inferior degree. The boat was slightly stove. The activity and skill of the lancers soon overcame this designing whale, accomplished its capture, and added its produce to the cargo of the ship."

Such intentional mischief on the part of a whale as seems to have been displayed in this instance is not frequent. It is probable, indeed, that nothing properly deserving the name of an intention to inflict injury can justly be attributed to the animal in any circumstances; these violent movements are merely the convulsions either of agony or of trepidation and intense fear. With all its enormous physical strength, the whale is singularly gentle and harmless—so remarkably so indeed that it has been characterized by those who have had the best opportunities of observing it as a stupid animal. It would require better proof, I think, than the mere absence of ferocity to make out this conclusion. There are some circumstances which would rather seem to show that the creature is possessed of considerable sagacity. It exhibits the usual instinctive sense of danger when it perceives the approach of its natural enemy, man; and, both before and after it has been struck with the harpoon, it most commonly adopts the very best expedients open to it to give itself a chance of escape. If a

field of ice be near, for instance, it makes for the water
under it, whither it cannot be followed by the boat; and,
even when it tries to release itself merely by a precipitate plunge downwards into the sea, it would be difficult
to say how it could act more wisely with a view to snap
the line to which it has got attached. If the effort were
not met on the part of the crew in the boat with the
most energetic application of those various resources of
art, dexterity, and decision, which are peculiarly at the
command of man, it would probably be in every case
successful. If it be the fact, also, as is asserted, that
the whales of the North Seas have abandoned certain
parts of their original domain, which are more accessible
to the fishing-vessels, and retired to other situations
which are more difficult of approach, this would seem
to imply, not only something of reflection and con-
trivance in individuals, but almost the possession of a
power in the species to transmit the results of experience from one generation to another.

The whale, in attempting to escape, sometimes exerts
prodigious strength, inflicting upon its pursuers not only
danger, but likewise the loss of their property. In 1812,
a boat's crew belonging to the Resolution, of Whitby,
struck a whale on the margin of a floe. Supported by
a second boat, they felt much at their ease, there being
scarcely an instance in which the assistance of a third was
required in such circumstances. Soon, however, a
signal was made for more line, and, as Mr. Scoresby was
pushing with his utmost speed, four oars were raised as
signals of great distress. The boat was now seen with
its bow on a level with the water, whilst the harpooner, from the friction of the line, was enveloped in
smoke. At length, when the relief was within a few
hundred yards, the crew were seen to throw their jackets

upon the nearest ice, and then leap into the sea; after
which the boat rose into the air, and, making a majestic
curve, disappeared beneath the waters, with all the
line attached to it. The crew were saved. A vigorous
pursuit was immediately commenced; and the whale,
being traced through narrow and intricate channels, was
discovered considerably to the eastward, when three
harpoons were darted at him. The lines of two other
boats were then sent out, when, by an accidental entanglement, they broke, and enabled the whale to carry
off in all about four miles of rope, which, with the boat,
were valued at 150*l.* The daring whalers again gave
chace; the whale was seen, but missed. A third time
it appeared, and was reached; two more harpoons were
struck, and the animal, being well plied with lances,
became completely exhausted and yielded to its fate.
By this time it had drawn out 10,440 yards, or 31,320
feet, making altogether about six miles of line. Un-
luckily, through the disengagement of a harpoon, a boat
and thirteen lines (nearly two miles in length) were
never recovered.

Occasionally the whalers meet with agreeable surprises.
The crew of the ship Nautilus had captured a whale,
which being disentangled and drawn to the ship, some
of them were employed to haul in the line. Suddenly
they felt it pulled away, as if by another whale, and,
having made signals for more line, were soon satisfied,
by the continued movements, that this was the case.
At length a large one rose close to them, and was quickly
killed. It then proved that the animal, while moving
through the waters, had received the rope into its open
mouth, and, struck by the unusual sensation, held it fast
between its jaws, and thus easily became the prey of its
enemy. The ship Prince of Brazils, of Hull, struck a

small whale, which sunk, apparently dead. The crew applied all their strength to heave it up; but sudden and violent jerks on the line convinced them that it was still alive. They persevered, and at length brought up two whales in succession, one of which had many turns of the rope wound round its body. Having become entangled under water, it had, in its attempt to escape, been more and more implicated, till, in the effort to escape, it ultimately shared the fate of its companion.*

The whale is one of those animals which once was considered as fit for the royal dishes, and we are informed that in ancient times, whenever one happened to be thrown on the British coast, the king and queen divided the spoil, the king asserting his right to the head, and her majesty to the tail.† And even at the present period it is customary not only with the sovereign of Great Britain, but I believe with all European monarchs, to claim any whale, or species of whale, that may be thrown on the coasts of their respective territories; and I believe this is confirmed to the English potentate by several acts of Parliament.‡

* Murray's " Narrative," &c. p. 401.
† " Blackstone's Commentaries," 1. c. 4. " Brit. Zool." &c.
‡ See the observations which conclude the account of the B. rórqual, in confirmation of this statement.

SPECIES II.

BALÆNA ICELANDICA, vel NORDCAPER; THE NORTH CAPE or ICELAND WHALE.*

THIS species differs from the preceding chiefly in having a more lengthened body, and a proportionally smaller head. It seems to agree with it in the absolute length of the body, which it sometimes exceeds. Its head has the appearance of an oval, truncated posteriorly and a little slanting at the extremity of the muzzle. The lower part is rounded and very high, and larger in proportion than the *B. mysticètus*. Amongst the drawings made by Mr. Backstrom, which the Count La Cépède has engraved, there is one which exhibits in a particular manner that oval appearance, which is maintained by the two bones of the lower jaw; these are united anteriorly, binding the acute extremities, and terminate in two processes, one of which is articulated with arm-bones (*ossa brachii*), forming a perfect oval. The *tout-ensemble* of the head, and of the baleen, are smaller than in the *B. mysticète*, when contrasted with its whole length. The dimensions of this species are much inferior to those of the common whale, and, as it is not much loaded with fat, the whalers seldom find it to yield more than on an average thirty tons of oil.

The two spiracles or air-holes represent two small semicircles, which are a little separated from each other, the convexities of which are opposed. The eye is very small, and its shortest diameter is placed obliquely.

---

* SYNONYMES. — *Balæna Mysticètus*, Var. B. Linnæus. *Baleine Nordcaper*, L'Abbe Bonnaterre, "Descrip. D'Island." La Cépède "Hist. Nat. De Cetacées." *Balæ'na Glacialis*, Klein. *Nordcaper*, Anderson's "Iceland." *Balæ'na Icelandica*, Dewhurst.

The edges of the baleen, which touch the tongue, are surrounded by black bristles, which act as a preservative from injuries it might sustain. The portion of the baleen touching the inferior lip is smooth and soft, but devoid of bristles or fringe.

The length of each pectoral fin exceeds the fifth part of its whole length, and its two arm-bones, which are contained within the fin, these are situated within the first division of its length.

The tail is slender at its extremity, where it forms a species of double fin, not only slanting, but as it were a little festooned; posteriorly the lobes are so long that from the external termination of one to the other there is an equal distance from the centre. On the belly of the male is seen a longitudinal fissure, the length of which is equal to the sixth part of the total length of the animal, the edges of which divide, so as to allow the actions of the organs of procreation. The anal orifice is a little circular opening, situated a little posterior to this longitudinal fissure.

The general colour of the *B. Icelandica* is of a gray, more or less distinct in its shades, which are rather uniform in their appearance. The lower part of the head often appears like a great oval of very shining white, at the centre and circumference of which are seen gray or black spots, irregular and confused.

Although the quickness of the B. mysticètus is surprising, that of the B. Icelandica is still more so; its tail being more slender and also flexible; its caudal fin more extended in proportion to its body, and this being also very flexible, affords it a much larger sweep when used as an oar, which is much quicker agitated; it is more powerful, and the strength with which it can flee, when pursued, darting like an electric shock

through the water, a great quantity of which it dis-
places.

This extreme velocity in the swimming of the Balæ'na
Nordcaper will on reflection appear highly necessary for
the purpose of its procuring food, inasmuch as this
species is not content like the B. mysticètus with feeding
on the *molluscæ, cancri,* &c., or on those animals which
are deprived of the powers of progressive motion, or are
unable to change their place but with great slowness.
The prey of this species, on the contrary, is extremely
rapid, from its preferring the Shad, and different species
of the mackerel, cod, and herring;* when it attains the
shoals of any of these fishes, or the banks of the shores,
it violently beats the water with its tail, creating a great
foam on the surface for some distance round. This is
done so quickly that the fish whom it intends to devour
are (according to La Cépède) rendered giddy, and, becoming paralysed from fright, can no longer oppose
themselves to its voracity, instantaneously become its
victims.   Mr. Willoughby counted thirty-two in a specimen he examined, which according to Martin had been
captured in Iceland; and in the stomach of another there
were discovered more than a ton of herrings.   Horre-
bow mentions that in one of these species the fishermen
of Iceland discovered no less than six hundred cod
fishes (many of whom were alive); and a great quantity
of shads and herrings were found in one that had been
thrown on the shores when pursuing his prey with too
great eagerness.† From these accounts, it appears that
this is one of the most voracious creatures inhabiting the
Arctic Ocean.

---

* The Zoology of these I shall describe in a subsequent part of this
work.

† Hist. Nat. de Cétacées, Par. M. I.e Comte La Cépède.

The shad, mackerel, and cod species are not un-frequently revenged by the *xyphias gladius*, or *sword-fish*, which forms one of the most formidable enemies to every species of the whale tribe; it however attacks this species with greater boldness, and becomes a despe-rate opponent, in spite of the velocity of his motions. We are informed by Martin that he once had an op-portunity of witnessing a most dreadful encounter be-tween the sword-fish and the Iceland or North Cape whale, being unable to approach with his vessel near to the infuriated combatants, who were endeavouring to effect the mutual destruction of each other. Having watched them for a considerable length of time pursuing one another, one precipitated itself on the other with great violence, at the same time inflicting such dreadful blows that the sea was thrown to an immense height around, and the spray falling down with the appearance of a thick fog.

In consequence of the quickness and agility of this whale, and being at the same time extremely savage, it becomes very difficult to encounter, and still more so to capture. However, when the whalers have been unsuccessful with the mysticètus, they then endeavour to destroy this species; for the capture of which they are compelled to employ a greater number of boats and harpooners, who must if possible be more active and alert than for the capture of the common whale. When this creature lies on the surface of the ocean, it shows but very little of its body and head above water.

The female Iceland whale is more easily captured either during pregnancy, or when nourishing her off-spring, her affection for which never allows her to for-sake it. When the whalers have been able to approach one, they must use great precaution, for it turns and

THE NATURAL HISTORY OF

returns with extreme force, rebounding and elevating
its caudal fins, and, feeling its danger, becomes dread-
fully furious, attacking the nearest boats, and with a
single blow of the tail will sometimes destroy them;
or, if it is compelled to yield to superior physical
strength, it then endeavours to escape, dragging with
it the harpoons, which, when successfully transfixed,
it will draw even a thousand fathoms of thick cordage
after it; and in spite of this immense weight, which is as
inconvenient as it is heavy, it swims with such rapidity
that the sailors in the boats can scarcely support themselves, without feeling a sense of suffocation.

The Norwegians run less risk in capturing this animal;
for, when it gets into the little bays terminating in a large
lake in their coasts, they then enclose the lake by nets
formed of the rind of trees, thus destroying the whale
without having the trouble and danger of fighting with
it.*

Duhamel du Monçeau informs us he was assured
that the blubber from the Nord Caper was devoid of the
disagreeable properties which have been attributed to
that from the balæna mysticètus.

---

* Sir Arthur de Capell Brooke states that he saw the skeleton of a
whale on the top of a stupendous high mountain called "Sandhorn," near
Gilleshall, in Norway, the height of which exceeds 3000 feet, and the
south side of it descends nearly perpendicular to the sea. In all probability this Zootomical specimen has remained there since the period of
the deluge, when it was deposited, which is now more than 4000 years!
No other conjecture I think can be formed than this, and it is wonderful
that these bones should have remained for such a period of time.—The
marine remains in the neighbourhood confirm this supposition.—The
species is not ascertained, but from its locality I should suppose it to be
the Nord Caper. *Vide* "Travels in Norway," &c. 237.

" On the top of the Fugelöe Mountain, according to the accounts of the
fowlers, who had often seen it, are the remains of a whale, lying in the
same manner as on the mountain of Sandhorn." *Ibid.* p. 331.

Dr. Klein has distinguished two varieties of this species, one he paradoxically names the " Southern Nord Caper," the back of which is very flat; and the other, whose back is more convex, he has given it the cognomen of the " Western Nord Caper," numbers of which I had opportunities of seeing.

The water ejected by this species is thrown out in the form of *radii*, and not in that of a *jet d'eau*, as is the case with the other whales, a fact with which the reader no doubt is familiar.

The varieties just mentioned, yet remain to be proved whether they are distinct, or whether they are mere sexual characters, or marks of age, or arising from unknown causes.

This species of the Cetàcea inhabits the Northern part of the Atlantic Ocean, situated between Spitzbergen, Norway, and Iceland. It also is a resident of the Greenland and Arctic Seas, where Mr. Backstrom made a drawing of one in 1779, and presented it to the late Sir Joseph Banks, P.R.S. It has also been met with in the Japan Seas, consequently in the Great Boreal Ocean in the 40th degree of latitude.

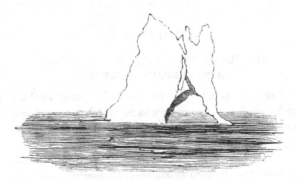

AN ICEBERG IN LATITUDE 76° 22′ N.

GENUS II.

THE BALÆNO'PTERÆ, OR FINNED WHALES.

THE animals which constitute this tribe agree with the true balæ'na in having their upper jaw furnished with horny laminæ being entirely destitute of teeth, and having two distinct orifices for the spiracles (or blowholes), placed towards the middle of the upper part of the head. But they are distinguished from them in being provided with a dorsal fin. Count La Cépède very properly divides this genus into two sections. The first of these is distinguished by the throat and belly being destitute of folds. The second has longitudinal folds beneath the throat and belly. According to La Cépède there are four species, one belonging to the first subdivision, and three to the second. I may observe that the word balænóptera is derived from balæ'na a whale, and pteron a fin or wing.

SUBDIVISION I.

SPECIES I.—THE BALÆNO'PTERA GIBBAR, OR RAZOR-BACKED WHALE.*

THIS species of whale is considerably longer and more slender than the common whale, which is probably the most bulky, and at the same time the most powerful, of all created beings.

---

* SYNONYMES.—Balæ'na Physalis. Linnè " System:" Gmelin ; Baleine Gibbar, Bonnat. " Encycloped. Methodique ;" Fin-Fish, Martin's " Spitz-

The principal characteristics of this animal, contrasted with the *B. mysticètus*, are the following:—First, as already mentioned, its body is longer and more slender; secondly, the baleen, or whalebone, is shorter; thirdly, it produces, when killed, less blubber and oil; fourthly, its colour has a much bluer tinge; fifthly, it has a greater number of *fins;* sixthly, in respiration, or blowing, it is more violent; seventhly, we find it quicker and more restless in its actions; and it is likewise considerably bolder in its conduct.

The greatest circumference of the *balænóptera gibbar* is from twenty-eight to thirty-seven feet, and its entire length more or less about a hundred feet. The longest lamina of baleen taken from this species measures somewhat about four feet. The greatest quantity of blubber recorded, of having been flensed from one of these animals was not more than ten or twelve tons. In colour, it bears a considerable resemblance to the sucking *mysticètus.* The body is not cylindrical, but is considerably compressed upon the sides, and angular at the back. A transverse section near the fins is an oblong, and, at the rump, a rhombus.

The respiration or blowing of this animal is very violent, and in calm weather may be heard at the distance of a mile. It swims with greater velocity than the common whale; and it has been estimated by Captain Scoresby to be about twelve miles an hour. It does not appear to be a very mischievous animal; neither is it timid or cowardly: when closely pursued by the boats, it exhibits very little fear, and does not attempt to outstrip them in the race, but merely endeavours to avoid them,

bergen ;" Pennant's "Brit. Zoology," vol. iii.; *Balænóptera Gibbar*, La Cépède, p. 114.

by diving or changing its direction. If harpooned, or
otherwise wounded, it then exerts all its energies, and
escapes with its utmost velocity; but shows very little
disposition to retaliate on its enemies, or to repel their
attacks by engaging in combat. The seamen frequently
endeavour, as an amusement, to wound it with a musket-
ball; but I never saw one that showed any symptoms
of injury from its effects. At a distance, it is occasionally mistaken for the *balæna mysticètus*; yet it is generally distinguished by its appearance and actions, which
are so very different. It seldom lies quiet on the surface
when blowing, but usually has a velocity of four or five
miles an hour: when it descends, it very rarely throws
its tail in the air, as is the case with the balæ'na mysticètus.

Besides the two half pectoral fins, it has a small
horny protuberance, which is a species of rayless or
immovable fin, at the extremity of the back; hence the
derivation of *razor-back* by the whalers.

From the small quantity of oil afforded by this animal,
and the great velocity with which it darts through the
water, renders it almost unworthy the time and attention
of the whalers. However, when it is struck, it not unfrequently drags the fast-boat with tremendous speed
through the water, but it is liable to be carried immediately out of the reach of assistance, and thus soon out
of sight of both boats and ships: hence, the harpooner is
commonly under the necessity of cutting the line, and
sacrificing his employer's property, for the purpose of
saving his own life, and those of his companions.

Captain Scoresby informs us that he has made several
attempts to capture one of these formidable creatures.
In the year 1818, this gentleman ordered a general chase
of them, providing against the danger of having his crew

separated from the ship, by appointing a rendezvous on
the shore, not far distant, and preparing against the loss
of much line, by dividing it at two hundred fathoms
from the harpoon, and affixing a buoy at the end of it.
Thus arranged, one of these creatures was shot, and
another struck. The former dived with such considerable
impetuosity, that the line was broken by the resistance
of the buoy as soon as it was thrown into the water; and
the latter was liberated within a minute by the division
of the line, occasioned, as was supposed, by its friction
against the dorsal fin. Both of them escaped. Another
B. gibbar was, however, harpooned by one of Captain
Scoresby's inexperienced harpooners, who mistook it for
a B. mysticètus. It dived obliquely with such velocity
that four hundred and eighty fathoms of line were
withdrawn from the boat in about a minute of time.
In consequence of the line breaking, this whale was also
lost.

The B. gibbar is found in great numbers in the Polar
Seas, particularly along the margin of those immense
fields of ice between Cherie Island and Nova Zembla;
likewise near the island of Jan Mayen. The Archangel
traders frequently mistake it for the *balæna mysticètus*.
It is rarely found amongst much ice, and appears to be
carefully avoided by the common whale; accordingly,
when seen by the whalers, it is viewed with painful concern. It inhabits chiefly the Spitzbergen quarter, the
parallels of seventy or seventy-six degrees; but in the
month of June, July, and August, when the sea is
usually open, it advances along the land to the northward as high as the eightieth degree of north latitude.
In open seasons, it is seen near the Headlands, at an
earlier period. A whale, probably of this species, measuring a hundred and one feet in length, was stranded

on the banks of the River Humber, about the middle of
September, 1750. The length of one of these whales
found dead in an inlet in Davis's Straits measured 105
feet, and its greatest circumference somewhat about
thirty-eight. The head was but small, when considered
in comparison with the common whale: the fins was
long and narrow, the tail being in breadth about twelve
feet, and was very finely formed. The baleen measured
four feet in length; was thick in substance, but bristly
and narrow as regards its breadth. The blubber on the
surface of the body generally is about six or eight inches
thick, and of a very indifferent quality. This species of
whale is usually of a bluish-black colour on the back;
and, on the abdomen, of a bluish-gray; the skin is
smooth, with the exception of the chest at the sides,
where there are several longitudinal rugæ or folds.

In the Zoological Museum at Bremen there is a
splendid skeleton of this species of whale. Professor
Peter Camper has ably illustrated the osseous structure
of this specimen, from which the annexed engraving of
the skull has been copied.

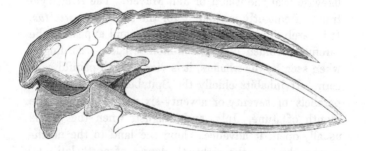

CRANIUM OF THE B. GIBBAR.

SUBDIVISION II.

SPECIES II. THE BALÆNO'PTERA ACUTO-ROSTRATA,

OR

SHARP-NOSED WHALE.*

THE *Balænóptera Acuto-Rostrata* is said to inhabit
principally the Norwegian Seas, and to grow to the length
of twenty-five feet.† One of this species was killed by
Captain Scoresby in 1813. The baleen (some of which
he has preserved) is thin, fibrous, of a yellowish-white
colour, semi-transparent, and almost like lantern-
horns: it is curved like a scymitar; and fringed with
white hair on the convex edge and point. Its length is
nine inches, and its greatest breadth two inches and a
quarter. One of these whales ran itself ashore on the
banks of the Forth, a little way above the town of Alloa,
on Sunday morning, October 23, 1808. About the break
of day, the servants of about the farm-house of Longcarse
were alarmed by the noise of the animal's blowing, and
floundering among the sludge. Assistance was speedily
procured from Alloa, when it was killed and secured.

As soon as Mr. Patrick Neill, the secretary of the
Wernerian Society, heard of the occurrence, he set off to
see the animal; but it was the 1st of November by the
time he arrived. By this time the flensing was over, the
blubber had been totally removed, and the kreng, or carcass, had been sent off to float with the tide, on account
of its offensive smell. However, he found it on the beach
at the village of Lower Airth, about two miles below

---

* SYNONYMES.—*Balænóptera Acuto-Rostrata* of La Cépède; *Balæ'na
Rostrata* of Linnæus and Fabricius; or the *Beaked Whale* of the Whalers.

† Count La Cépède states the length at eight or nine metres, which is
twenty-six to twenty-nine feet.

H

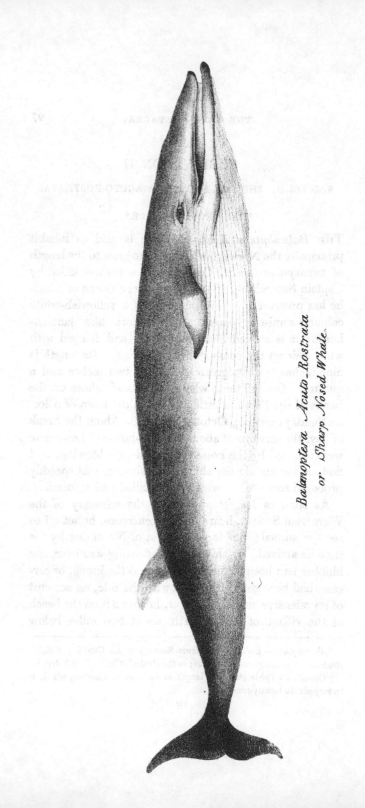

Balænoptera Acuto-Rostrata
or *Sharp Nosed Whale.*

Alloa, and on the opposite side of the river. The total length of this animal was forty three feet; its circumference, where it was thickest, immediately behind the swimming paws, was about twenty feet. It had a single dorsal fin (if the horny protuberance on the back may thus be called) two feet six inches high, and nearly the same in breadth at the base—very diminutive, certainly, when compared with the bulk of the animal's body. This fin was placed very far down the back, about twelve feet only from the extremity of the tail, and nearly over the vent: it was of an acute triangular shape, blunt in front, and sloping off to a thin edge behind, slightly hooked, the curvature being towards the tail. From its shape, the sailors engaged in whaling call it a pike. The under jaw projected about three inches beyond the upper; it was nearly fourteen feet long, and somewhat broader or wider than the upper jaw.

In the upper jaw, there were two rows of short baleen, the laminæ of which were placed perpendicularly, and very closely set together: the largest were in the middle of the rows, and were only about eighteen inches long. Each lamina was dark-coloured in the thickest part, but became of a greenish or bluish-white colour on the thin side, next the interior of the mouth, where it separated into white hair or bristles : there might be about three hundred on each side.

The tongue was black, of great size, soft, and nearly smooth.

There were two blow-holes, long and narrow apertures separated from each other only by a thin partition. They were situated in the highest part of the upper jaw. The eyes were placed on the sides of the head, a very little way behind and above each angle or corner of the mouth. From eye to eye, measuring across the head, was nearly seven feet. The socket of the eye was fully two inches and a half in diameter.

The skin was black on the back; but towards the belly the colour changed to whitish. The cuticle was very fine; the true skin, soft, spongy, and of considerable thickness.

The whole skin of the thorax and upper part of the belly, was plaited or folded. The *sulci*, or plicæ, as Sir Robert Sibbald calls them, were about two dozen in number. They extended from the lower lip to about four feet beyond the swimming-paws. On the under jaw, they ran obliquely downwards; but on the belly they had a straight longitudinal direction; on the fore part of the body they were uniform and parallel, but diverged a little towards their termination behind. The *flencers*, having found little or no blubber under the plaited skin, had left a considerable portion of it untouched.*

The back was rounded next to the head; a little before the dorsal fin, it began to assume a somewhat angular shape, and this form was continued till a subordinate short ridge marked the commencement of the tail. The flattened or extended part of this member was, as in other species, horizontal, and divided into two lobes. The breadth, measuring between the extremities of the lobes, was no less than ten feet; and its depth was nearly three feet.

The swimming-paws, measuring from the tip to the

---

* The use of these folds in the skin of the thorax, which long was a problem, has now been ascertained. They are calculated to permit the animal to swell up a large pouch or bladder, placed in the anterior part of the body. When this bladder is expanded, the folds disappear, and the animal appears as if it was striped, the covered interstices of the folds being of a paler colour than the rest of the skin of the thorax.

arm-bone, or ball which is received into the cavity of the scapula or shoulder-blade, were nearly five feet long. In breadth, at the widest part, they did not extend to one foot, tapering to a pretty sharp point. They were narrow at their junction with the body. The socket that received them was large, being four inches and three quarters in diameter.

The dorsal or largest vertebræ were eight inches in diameter. None of the others were laid open so as to admit of examination. The animal was of the male sex.

The blubber was firm in its texture, and not unlike pork fat, when softened by heat. It does not appear to be so subject to putrescency as the blubber of the common whale. It filled seven large casks, but was not expected to yield much oil, compactness not being a desirable quality in blubber. A soap-maker bought the whole for about 15l. sterling.*

The late John Hunter dissected one which was caught on the Doggerbank: it was seventeen feet long.†

Captain Scoresby does not consider the one described by Mr. Neill to be a B. Rostrata, but rather a Balænóptera Jubartes, or to an undescribed species.

Another caught in Scalpa Bay, in November, 1808, was seventeen feet and a half long, circumference twenty. Length from the snout to the dorsal fin, twelve feet and a half; from the snout to the pectoral fin, five feet; from the same to the eye, three feet and a half; and from the snout to the blow-holes, three feet. Pectoral fins, two feet long, and seven inches broad; dorsal fin, fifteen inches long by nine inches high; tail, fifteen inches long by four feet and a half broad. Largest piece of baleen,

---

* *Vide* Mr. P. Neill's paper in the "Memoirs of the Wernerian Society of Edinburgh," vol. i. p. 201—6.

† Philos. Trans., vol. lxxvii. p. 448.

about six inches : colour of the back, black; of the belly, glossy white ; and of the grooves of the plicæ, according to Dr. Traill,* who saw it in Scalpa Bay, a species of flesh colour.

SPECIES III.

BALÆNO'PTERA JUBARTES, THE PIKE HEADED OR FINNED WHALE.

THE pike-headed whale or finner† is seen traversing the ocean between Newfoundland and the British Islands, in great numbers; but, during the months when the common whale (B. mysticètus) sallies forth from his haunts, they are observed running towards the Arctic Seas, and are considered good guides to the retreats of the Greenland whale. Like that animal, the place of teeth is supplied by a quantity of baleen or whalebone in the interior of the mouth, firmly imbedded in the frontal bone ; but in this animal they are shorter, and of a bluish colour.

The finner is also much thinner in blubber, more slender, and measures about forty-six feet, and in circumference of body about twenty feet. The dorsal protuberance or fin is somewhat about two feet and a half high ; the pectoral fins vary from four to five feet long, externally, and scarcely a foot broad, the tail about three

---

* This Gentleman is now one of the Professors of the University of Edinburgh.

† SYNONYMES.—*Balæ'na Boops.* In France, the *Balænóptera Jubartes.* In Greenland, the *Keporkak.* In Iceland, the *Hrafu-Reydus.* The *Finner*, by the Whalers, and the *Pike-Headed Whale*, by Pennant.

Sir Arthur de Capell Brooke is disposed to consider this and the B. Nordcaper the one and same species, from their general locality; the dorsal fin, however, marks the difference.

feet deep and ten broad; of the baleen, there are about three hundred laminæ on each side, the longest of which is about eighteen inches in length; the lower measures nearly fifteen feet, or one-third of the whole length of the animal; there are about two dozen sulci, and two external blow-holes or spiracles; the blubber measured on the body only two or three inches in thickness, but under the sulci none. This whale lives principally upon a small species of salmon, which is denominated the *salmo arcticus*, as well as on the *argonautica arctica*,* and the *ammodytes tobianses*† or lance. When in the act of opening its mouth, it dilates the abdominal plaits or furrows, which lie in pairs, and, on account of the colour of their internal surface, present at this juncture a beautiful spectacle; the fore part of the belly as if it was elegantly striped with red.‡

According to Mr. O'Reilly, the finner is gregarious, being found usually in herds from five to a dozen, and they are at any distance easily to be distinguished by the strength, elevation, and whiteness of the watery column issuing from their blow-holes or spiracles. The blast of the finner is forced directly upwards in a firm column of more than ten feet, and with such an accompanying

---

* This is a species of the *sepia* or *clio*, having a univalve shell, which is spiral, involuted, membranaceous, and containing only one cell. "Rees's Cyclopedia," vol. 2.

† This forms a genus in the Linnean system, only one species of which has heen hitherto discovered. The generic character is as follows: the head compressed, narrower than the body; upper lip doubled, lower jaw narrow and pointed; sharp-pointed teeth; gill membrane of seven rays; body long and square; tail fin distinct. This species is the tobianses of Linnæus.—It inhabits the sandy shores of the Arctic Seas: it is usually from nine to twelve inches in length, and of a silvery white colour, with a greenish back. Rees, vol. 2.

‡ Shaw's "Zoology." vol. 2., part. 2., p. 493.

gust as in a calm evening to be heard at the distance of more than half a mile. The attention of the whale seamen is drawn to the path of the finner by the noise of this discharge; and should the animal lie beneath the surface, and his course be tracked by the eddying ripple caused by the motion of the whale, which differs a little from that of the common mysticète, and if the harpooner ascertains it to be the finner's blast, he immediately suspends all operations for effecting its capture.

According to Sir Arthur Brooke, this whale is found on every part of the Norwegian coast, in great numbers. They even make their appearance in the harbour of Hundholm, and a few days before that gentleman's arrival a very large one was seen by the sailors close under the bows of the ship Eliza, which was then loading with stockfish (dried cod) for the Mediterranean.* It is likewise seen in great numbers about Tromsöe and Hvalöen. During Sir Arthur's voyage up the coast, he had repeatedly heard from the fishermen of the mischief they occasionally do, and the danger there is in meeting with them, particularly during the months of July and August, when they collect together, and, if any boat comes in their way, it runs a risk of being upset. When he was near three hundred miles farther south, a report had reached those parts of the damage occasioned by some sea animal between Bodöe and Tromsöe; and this was now found to have been a whale, which had inspired such an alarm. The foregoing year (1819) so much fear of real or imaginary danger had been excited that the communication between the north and south had actually for a short time been stopped, as no fisherman would venture out, through dread of meeting with the

---

* Travels in Norway, &c., p. 254.

finners: and when the English brig St. John, which was
bound for Hundholm, was lost off the islands of Röst, at
the extremity of the Lofodens, no money could induce
the fishermen to carry a letter with the intelligence, from
the dread of having their boat upset by the whales. The
alarm arose from a whale, which was described with a
large black tuft of hair upon its forehead,* having pursued two boats near Tromsöe, one of which it dashed to
pieces, and the other escaped with considerable difficulty
by running ashore.

The circumstance of the whale's pursuing boats in the
months of July and August has been mentioned by
former travellers, and, though singular, Sir Arthur is disposed to give credence to it, from its having been confirmed by many of the respectable inhabitants in Nordland and Finmark.

The fishermen suppose these to be the males, and that
the reason of their pursuing the boats during the hottest
months is their mistaking them for the female; and it
is from their rude embraces that the boats suffer so
greatly. From the note† below, however, which gives

---

* I suspect they meant at the extremity of the muzzle, which is characteristic of the *B. Rórqual*.

† " The American ship the Essex, G. Pollard, Captain, was on a whaling voyage, and in the latitude of 47 degrees south, and longitude 118
west, when the following accident happened. They were surrounded
by whales, the three boats were lowered down, and the crews busy in
harpooning them. Shortly afterwards a whale of the largest class struck
the ship, and knocked part of the false keel off. The animal then remained
some time alongside, endeavouring to clasp the ship with his jaws, but
could not accomplish it. He then turned, went round the stern, and, going
away about a quarter of a mile, suddenly turned, and came at the ship with
tremendous velocity, head on. The vessel was going at the rate of five
knots, but such was the force when he struck the ship, which was just
under the cat-head, that the vessel had sternway (went back) at the rate
of three or four knots. The consequence was, that the sea rushed in at

an account of an extraordinary attack made upon a vessel by a whale, it will be seen that in this instance the animal must have been instigated by revenge. It is an extract from a New South Wales paper, and the consequences were singularly fatal as well as horrible in their nature.

To show still further the danger that a vessel may be placed in, even from accidentally encountering a whale, Von Longdorff, in the narrative of his voyage from Kamtschatka to Ochotsk, says, " an uncommonly large whale, the body of which was larger than the ship itself, lay almost at the surface of the water, but was not perceived by any one on board, till the moment when the ship was almost upon him, so that it was impossible to prevent its striking against him. We were thus placed in imminent danger, as this gigantic creature, setting up its back, raised the ship at least three feet out of the water. The masts reeled, and the sails fell all together, while we who were below all sprang instantly upon the deck, concluding we had struck upon some rock. In-

her cabin windows, every man on deck was knocked down, and the bows were completely stove in. In a few minutes the vessel filled, and went on her beam ends. At this unhappy juncture the captain and mate were each fast to a whale ; but, on beholding the awful catastrophe that had taken place, they cut immediately their lines, and made for the ship. The boats were prepared ; a small quantity of bread and water quickly put on board, and the crew got in, part of whom, after having been ninety days at sea, were providentially picked up by another whaler and saved. The horrible sufferings they had experienced were no less dreadful than the extraordinary circumstances which had occasioned the loss of their ship. Their bread being consumed, they were under the dreadful necessity of casting lots, to determine who should die, to afford sustenance to the rest. Eight times had lots been drawn, and eight human beings had thus been sacrificed, when a vessel encountered them. At that time the captain and a boy had also drawn lots, and it had been determined that the latter should die, when he was thus unexpectedly saved." *New South Wales Paper.*

stead of this, we saw the monster sailing off with the utmost gravity and solemnity."

A very singular circumstance, as related to Sir Arthur Brooke, in the Nordlands and Finmark, is the partiality these enormous animals have for cows or horses; and he was acquainted with a merchant at Tromsöe who, having some of the former in a boat, was so constantly pursued by them that he was obliged to land and put the cows on shore. At Röst is a small inlet, or narrow creek, at the extremity of which are large cow-houses; and it happens that almost every year whales are taken in it, being attracted, it is said, by the smell of the cows or their ordure; when, not being able to return, they fall a prey to the fishermen.

The manner in which the whale-fishery is conducted about Tromsöe and other parts of the Finmark coast, by the Laplanders, is singular. When a fin whale is discovered, two of them go in pursuit of it in a small boat. On approaching it, as soon as they have succeeded in plunging the harpoon into the animal, they immediately break it off close, and their business is finished. They think nothing of farther securing the whale, which, with inconceivable velocity, makes off from its cruel enemies, but bearing in it the deadly mark of their attack plunged deep in its body, and in the course of a few days it is generally found dead on some part or other of the neighbouring coast. The person who finds it gives notice of it, and the fisherman who struck it comes and identifies his property, by his name or mark on the barb of the harpoon. The finder is then rewarded by one-third of the booty, to which he is by law entitled.*

The inhabitants of Kamtschatka make use of every

---

* Sir A. Brooke's Travels, p. 300.

portion of the B. jubartes. The oil serves them partly for fuel, the preparation of their food, and affords them light. The delicate pieces of baleen, or whalebone, they make into threads for the manufacture of fishing nets, lines, &c. The lower jaw-bones are used as portions of sledges, handles of instruments, &c.; sometimes the ribs form the frame-work of the cabins; the nerves answer the purpose of cord; and the various portions of the stomach and intestines form vessels to contain their drink and oil. The skin, which they rudely tan, they form into sandals, bags, and harness.

I may observe, in reference to the foregoing, that the harpoon does not generally kill the whale, at least the B. mysticètus; for one which I saw captured, in 1824, had a harpoon (nearly bent double) cut out from near the left fin, which had been struck by a harpooner of the ship Majestic of London,* Captain Lawson commander, in 1819, where it was firmly imbedded. The harpoon was brought home, and presented by Captain Ainslie to the owners of the vessel, Messrs. Benson and Hunter of Shadwell, London.

SPECIES IV.

BALÆNO'PTERA RO'RQUAL, or BROAD-NOSED WHALE.†

THE natural history of the species of whale, whose skeleton‡ I shall shortly proceed to describe, is as follows.

---

* This fine vessel was subsequently wrecked; it was the property of William Mellish, Esq.

† SYNONYMES.—*Balæ'na Musculus*, Linnæus. *Balænóptera Rórqual*, La Cépède. In Iceland it is called the *Steipe Rey-das*. The Broad-Nosed Whale of the whalers.

‡ In April 1822, I published a long article illustrative of the zoology

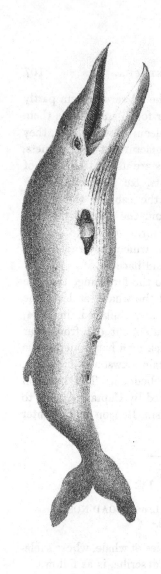

The Balænoptera Jubartes or Pike-Headed Whale.

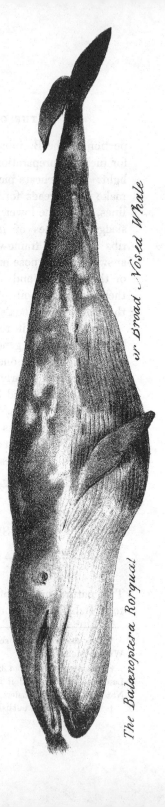

or Broad Nosed Whale.

The Balænoptera Rorqual

On the 4th of November, 1827, some fishermen of Ostend discovered the dead body of a female whale floating in the sea, between the coasts of England and Belgium. Not being able to tow the enormous carcass themselves, the master of the shallop Dolphin of Ostend, who had likewise discerned it, employed the aid of his vessel and crew to move it, but without success. They then called to their assistance two other vessels, and by their united efforts surmounted the difficulty, and were enabled to appear in sight of Ostend at four o'clock next day; as soon as they entered the harbour, the rope broke, and it was cast upon the eastern side.

The appearance of a whale of such enormous dimensions created a great sensation, inasmuch as those which had formerly been stranded or captured on the coast of Flanders were of much smaller dimensions, and none had appeared during the present century. These, however, I shall briefly notice.

In the year 1178, the magistrates of Bruges offered to Count Philip a sea monster, or whale, which had been thrown, in consequence of a great tempest, on the coast of Ostend. This animal measured forty-two feet in

---

of this animal, and of the anatomy of the skeleton, then exhibiting on the site of the late King's Mews at Charing Cross. This paper having been approved of by many eminent zoological anatomists and naturalists, I have at their request given it a place in this work; it is copied from the Magazine of Natural History, so ably conducted by J. C. Loudon, Esq. I may here observe, that the word *Rórqual* signifies, in the Norwegian language, "a whale with furrows;" hence it is very expressive of the distinguishing characteristics of this animal.

The lithographic drawing of the *B. Ròrqual* will be found on the same plate with the *B. Jubartes:* it was taken previous to the dissection of the animal, by order of Mons. Kessel at Ostend, in 1827, and is the only correct one extant; it is copied from the large one presented to me by that gentleman, in May 1832. The scientific reader is respectfully requested to contrast this view of the "Rórqual," with the inaccurate one given by Count La Cépède, and then remark the striking difference.

length. The formation of the mouth and head is recorded as bearing a resemblance to the beak of an eagle and the figure of a sword.*

The chronicles of Flanders report, that, in the month of November, 1402 or 1403, there were eight whales stranded before the port of Ostend, the longest of which measured nearly seventy feet, and produced nearly twenty-four tons of oil.† On the 20th of January, 1762, there was discovered a dead whale, measuring forty French feet in circumference, on the ride between Blankenberg and Ostend, nearer to the latter city. After having been exposed to the public for five days, it was sold for the benefit of the sovereign, for the sum of 192 Flanders florins (about 16l. 13s. 4d.) Several of these creatures have at different times been killed or stranded upon the British coast. Captain Scoresby has recorded several of these events. One was captured on the coast of Scotland, in the year 1692. Another was fifty-two feet long, and had been stranded near Eyemouth, on the 19th of June, 1752. Another, nearly seventy feet in length, ran ashore on the coast of Cornwall, on the 18th of June, 1797. Three were killed on the north-west coast of Ireland, in the year 1762, and two in 1763. One or two have been killed in the river Thames. Another was embayed and destroyed in Balta Sound, Shetland, in the winter of 1817-18, some of whose remains were seen by Captain Scoresby, Jun., who thus states its dimensions :—Length, eighty-two feet; lower jaw-bones, twenty-one feet each; longest blade of the baleen or whalebone, about three feet. Instead of hair at the inner edge and point of each lamina,

---

* From this vague description, I am of opinion that it was a *Balænóptera Acuto-Rostrata*, or the sharp-nosed whale of Pennant.

† Dr. Dubar has not mentioned the species which they belonged to.

it had a fringe of bristly fibres, and was stiffer, harder, and more horny in its texture, than the same part in the common Greenland whale. The quantity of oil produced from the blubber of this animal was only about five tons, of very inferior quality, some of which was extremely viscid and bad. The total value of the whole, deducting all expenses of extracting the oil, &c., was no more than 60*l.* sterling. It had the usual sulci or furrows about the thorax or chest and dorsal fin.

To return to this rórqual: Mons. Herman Kessels of Ostend formed the idea of preserving so valuable an acquisition in zoology and comparative anatomy within the kingdom, instead of allowing it to be made a source of mere pecuniary profit. The perseverance, philanthropy, and enterprising spirit of this gentleman are well known. During the inclement winter of 1827, he contributed to the comforts, health, and happiness of thousands of the indigent of Ostend, by daily distributing food, soup, and warmth among them. To cover the great expenses of this benevolent act, he addressed himself to the wealthy of the town to further his beneficent design, which alone procured him the blessings of all who had tasted of his bounty. Mons. Kessels had scarcely formed the idea of preparing the skeleton of this whale ere it was commenced; as he publicly purchased it for the sum of 6230 francs (about 259*l.* 11*s.* 8*d.*), jointly with Dr. Dubar, an eminent physician of Ostend, on the 16th of November, 1827. From the time the rórqual was thrown into the harbour, considerable doubt was entertained in the minds of many scientific naturalists as to what species it belonged to; some declaring it a cachalot, others a gibbous whale, &c. &c. However, from its possessing the longitudinal folds extending from the throat towards the middle of the trunk, it was indicated to be either a rórqual or the finner of the whalers: the latter is the pike-headed whale of Pennant. Various reasons decided it to belong to the former species; but every work by professed naturalists exhibited contrary opinions. Even the illustrious Cuvier himself was in error, inasmuch as he states that all cetàcea with folds belong to one and the same species; whereas, according to Count La Cépède, the dorsal fin proved it to belong to the second class of the whale genus, which he has named Balænóptera.

Towards the end of November, 1827, M. Kessels went to Paris, where he consulted Baron Cuvier, and returned with Messrs. Dubar and Parét, the latter an eminent amateur naturalist, on the 20th of December. They had exhibited to this zoologist the whole of the drawings which had been taken of the animal; and he informed them that the Balænóptera *Rórqual* and the Balænóptera *Jubártes*, which La Cépède and other naturalists had described as two species, were only one and the same, as their distinguishing characters were so trifling that they might be easily confounded with each other. However, Dr. Dubar, notwithstanding this opinion, very properly determined on considering it a Rórqual in the pamphlet which he published on this subject. To whatever species the individual specimen in question belongs, it is doubtless the largest animal that has ever been captured, and I do not hesitate to say that the skeleton is the most perfect in Europe.

The following measurements will give the reader some idea of the bulk of this animal:—

Total length of animal, 95 feet; breadth, 18 feet. Length of the head, 22 feet; length of the lower jaw-bones each, 22 feet; height of the skull, 4½ feet. Length of the spine, 69½ feet; number of bones composing it,

54. Length of ribs, 9 feet; number (14 on each side), 28. Length of the fins, 12½ feet. Length of the fingers, 4½ feet. Width of the tail, 22½ feet; length of the tail, 3 feet.

Weight of the animal when found, 249 tons, or 480,000 pounds.

Weight of the skeleton, only 35 tons, or 70,000 pounds; being a little less than one-seventh of the entire bulk.

Quantity of oil extracted from the blubber, 4000 gallons, or 40,000 pounds.

Weight of the rotten flesh buried in the sand, 85 tons, or 170,000 pounds.

The dissection of this animal commenced under the superintendence of Dr. Dubar, on the 14th of November, in the presence of a great number of medical and other scientific men. The workmen were sixty-two in number, who were employed both day and night; they constructed a wooden house close to the spot. By the 19th the skeleton was dissected out, and deposited in a place prepared for that purpose; but it was not until the 20th of April, 1828, that it was articulated, and fit for exhibition. For this purpose the carpenters commenced on the 14th of January the construction of the pavilion for its reception, the same that was recently at Charing Cross, of which the following engraving is a representation. I may observe, that the building displays very considerable ingenuity on the part of its contrivers; as it is so constructed, that it can be pulled to pieces, and re-erected at a distant spot, in a few hours when required.

PAVILION OF THE GIGANTIC WHALE.*.

When it was completed, Mons. Kessels, with the greatest liberality, gave several grand entertainments to the scientific men of the town, as well as to the workmen who had been employed, and likewise to the poor of the town; in fact, there were several days of great rejoicings. Medals of gold were presented from Mons. Kessels, by the governor, the burgomaster, and by Lieutenant-colonel Dufrenery, commandant of the place, to the heads of the following Societies:—To Mons. Jacques de Ridder, as president of the Royal Society of Saint Sebastien; to Mons. Philippe de Brock, president

---

* This was the inscription on the portico, immediately over the entrance.

I

of the Society of Saint Andrew; to Mons. Aimé Lie-
baert, president of the Royal Society of Rhetoric, who
also received from the same gentleman the fourth medal,
which had been offered as a prize to the musical department of the Society.

This genus is found not to remain so much to the
northward as the common Greenland whale (*B. mysticètus*), inasmuch as I have already stated its occasional
occurrence in the seas about Great Britain, Ireland,
Norway, and other nations near the Arctic Seas; it has
also been found in the Mediterranean, near the Straits
of Gibraltar. The proportion of oil which whales of
this genus and species furnish is not to be compared
with that supplied from the balæ'na mysticètus; and
the baleen, or whalebone, from its smallness, is not so
valuable. These circumstances, together with its great
velocity, make this species a matter of indifference to
the whalers, who rarely attempt its capture. This protuberance, in conjunction with a series of longitudinal
furrows from the throat to the anus, points out the individuals possessing them as either of the kind called pike-
headed whale, or rórquals. Both kinds are discovered
near the 75th degree of north latitude. The rórqual
subsists principally upon herrings and smaller fish, and
its consumption of these must be immense, when we
consider its vast size.

The back of this whale, when captured, was of a
blackish hue, and the belly white. The lower jaw is less
pointed than those of the other cetàcea, which is also a
distinguishing mark of this species. The eye is situated
near the opening of the posterior part of the lips; and
as the condyles (knobs which fit into sockets at joints) of
the lower jaw are very high, so that the top of the head
is almost on a level with the neck; the visual organs are

therefore so contiguous to the top of the head, that they frequently appear above the water, when the rórqual is swimming on the surface. The pectoral fins are placed at a short distance from the opening of the mouth, and nearly at right angles with the lips when extended. The dorsal fin is situated above the opening of the anus, and is very small in proportion to the size of the animal. The tail is divided into two lobes, with a convexity on the posterior portion of each; the inner margins of each lobe unite directly in the middle, in a line with the termination of the spine.

In its manner of blowing (*i. e.* respiration), swimming, and general actions, as well as its appearance in the water, it can hardly be distinguished from the *B. gibbar* during life, but on examination after death, it is found to be not only a much shorter animal, but possessing a larger head and mouth; with a rounder jaw than that animal.

OBSERVATIONS ON THE ANATOMY OF THE SKELETON.

Having given a brief outline of the zoological characteristics of this whale, I now proceed to make some observations on the anatomy of the skeleton,* which, as I have remarked above, is that of a female.

In this skeleton there are several anomalies by which it is rendered peculiar, when contrasted with the other mammàlia. There are but two distinct kinds of articulation, viz. first, the hinge kind, as in the articulation of the lower jaw with the head; and, secondly, the ball and socket kind, forming the joint of the shoulder, on the articulation of the arm-bone with the scapula, or

---

* The reader may rely on the accuracy of the annexed plate, which was taken from the skeleton under my superintendance, by an ingenious artist.

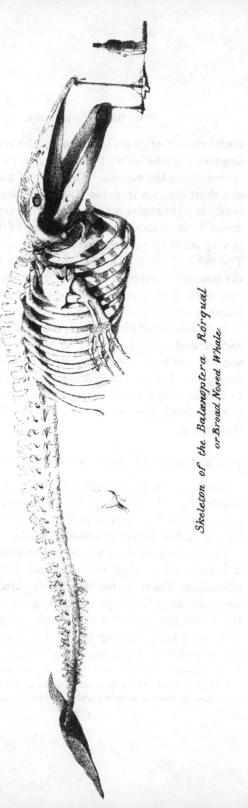

Skeleton of the Balænoptera Rórqual or Broad Nosed Whale.

shoulder-blade. There are none of the movable or the semi-movable articulations. Those I have mentioned possess cartilaginous surfaces, as they do in other animals; and thus the effects of friction are prevented: the other bones are only united by ligaments, which, however, do not form any capsules; they are interosseous, and serve more the purposes of agility than flexibility. The greatest portions of the skeleton are united through the medium of intervening cartilages, even to the fingers, that is, the bones within the pectoral fins. The sutures are imperfectly formed, and in some places the kind of suture termed harmonia can hardly be said to exist; in the head, especially, the union of the bones is so feeble, that they appear nearly disunited.

Most of the bones of these animals are very porous, and contain large quantities of very fine oil. The lower jaw-bones, which measure usually from twenty to twenty-five feet in length, are frequently preserved on account of the oil, which can be drained from them when they are conveyed into a warm climate. When this is exhausted, these bones float freely in water. They have very little of the compact substance which usually characterises bones, and in some parts form portions which are denominated *epiphyses*, that are but feebly connected to the other bones; and in the spine thirteen transverse natural processes were found detached from the body of the bone, without any apparent cause. Another peculiarity exists in the articulation of the ribs, which are not united to the bodies of the vertebræ, as in other mammàlia, but are connected through an intervening cartilage to the transverse processes of the dorsal vertebræ. This portion of the skeleton is pretty nearly solid.

According to the observations of the late Sir Charles

Giesecké, the balæ'na mysticètus, or common whale, possesses thirteen* ribs on each side; whilst in the B. rórqual there are fourteen. An additional distinguishing character in the rórqual is the circumstance of there being at the muzzle a few small blades of baleen, or whalebone, a character not found in any other species of the whale genus, with a small bristly tuft, like the mane of a horse, only much firmer in texture. This important feature in this animal is finely preserved in the skeleton. This important and characteristic fact has neither been mentioned by Pennant nor Cuvier, and the rórqual in La Cépède's " Histoire Naturelle de la Cetacées" is any thing but a true representation. There are no abdominal or hind limbs in any of these animals ; neither is there any vestige of pelvis, with the exception of a small portion of bone analogous to the ossa pubis of quadrupeds.

The Head.—This portion of the whale bears some resemblance to a pyramid lying on its side, the point or apex being in the front, and the base attached to the spine. We may not improperly divide, for the purpose of description, the head into five surfaces, viz. a superior, an inferior, a posterior, and two lateral. The superior surface is of a triangular shape; its length being about twenty-five feet: it is terminated anteriorly by the muzzle or extremity of the palatine bones ; and posteriorly by the vault of the skull, which is occupied by the brain, and is distinguished by the frontal bone, which, passing in a semilunar direction, terminates in a process that contributes to form the anterior portion of the zygomatic arch; thus exhibiting an analogy to quadrupeds. From the top and anterior part of the frontal bone the

---

* Vide, p. 45.

nasal bones are articulated by sutures, and extend the whole length of the upper part of the mouth. Beneath these are two ossa vomeres or ploughshare bones, forming two thin osseous laminæ, and these are closely connected to the inferior part of the frontal bone.

The superior surface of the palatine bones may be perceived externally, and they are of a more spongy texture than the preceding. They are of a triangular form, and are curved at the external margin. Towards the posterior part there are five or six large foramina or orifices, which afford a passage for the nutritive arteries, &c. Between the above bones there is a large space left in the upper part of the mouth, which affords a lodgment to the ethmoidal bone (a bone that, in the superior part of the human nose, is said to resemble a sieve); and also for the spiracles, through which the animal ejects water; and these are popularly denominated the blow-holes. The ethmoidal bone is placed in the cavity formed by the nasal bones, and by which it is concealed: it is light, spongy, and formed of thin laminæ.

The lateral surfaces of the skull are likewise nearly triangular, and extend superiorly only to the sides or parietes or walls of the nose; presenting several furrows which afford a lodgment to several important blood-vessels and nerves. The use of these bones is to augment the nasal cavity; they are lined by a dense, thick, olfactory membrane, in which the organ of smell is situated.

The inferior surface, like the rest, is triangular, and is mostly formed by the principal part of the palatine bones, and likewise possesses a great number of furrows and canals which afford a passage to the nutritive vessels and nerves; in the exterior boundary there is a sulcus or furrow, which indicates the place where the baleen or whalebone is inserted. At the posterior part of this surface, and between the mastoid processes,* the two bones containing the organ of hearing, denominated the petrous, or, as I term them, the acoustic bones, are placed. In the interior there is a nervous pulp, in which the sense of hearing is supposed to reside.

The posterior surface or base of the skull is of a semi-circular form, with two large alæ or wings on its sides, at the bottom of the pterygoid or wing-like processes of the sphenoidal or wedge-like bone. There are the humular or hook-like processes, to which the pharynx or upper part of the gullet is attached. The great occipital foramen or orifice in the back of the head for the passage of the spinal marrow from the brain, is situated a little above the preceding. On each side of this foramen or orifice there are the semilunar condyles or knobs of the occipital (hind head) bone, which are articulated with the atlas or the first bone of the neck as in the other mammàlia. The remainder of this portion of the skull is occupied by the greater part of the occipital and the mastoid processes.

The lateral surfaces are formed by the end of the palatine bones anteriorly; the zygomatic fossa or cavity and its arch posteriorly. This surface embraces portions of the temporal, occipital, and the sphenoidal bones.

On viewing the head vertically, we find several interesting peculiarities; the occipital bone measures more than three feet in thickness, and is very spongy in its texture, whilst the external table is at the same time extremely thin; consequently the specific gravity must be very little, notwithstanding its immense size. The nasal cavities are very largely developed, and in the

---

* Processes of the temporal bone, shaped like the nipple of the human female breast.

living animal not only contain the olfactory membrane, but likewise the spiracles, or organs by means of which the whale is enabled to project water to a considerable height above the surface of the ocean. The cerebral cavity, when contrasted with the dimensions of the other portions of the body, is extremely small; beneath it is the point of union of the vomer with the occipital and part of the ethmoidal bones. With the exception of the lower jaw-bones, all those composing the head are of a spongy nature, and appear to be formed of a series of laminæ. The lower jaw, like the same portions of other animals, and of the human infant at birth, is formed of two distinct pieces of bone, united together at the point or chin by symphysis, or a thin layer of intervening cartilage: each one forms a curve terminating in its condyle, and measures twenty-two feet in length from the chin to its articulation with the bones of the head. It is extremely hard and compact; the coronoid process which is separated from the condyles by an almost horizontal space, which occupies the place of a semilunar cavity found in the other mammàlia, affords insertion to the temporal muscle. They articulate themselves with the glenoid cavities of the occipital bone, in such a manner as to form a perfect hinge joint. The superior margins of these bones are perfectly smooth, and exhibit not the slightest vestige of any alveolar cavities for teeth, which are found in several genera of the order cetàcea.

There is a number of large foramina on the labial surface of these bones, for the passage of large blood-vessels. The anterior mental foramen is placed externally near the chin, and is sufficiently large to admit a man's thumb: this leads to a large canal, which traverses the body of the bone; it contains blood-vessels and nerves, which, having performed their important duties in nourishing the bone, pass out by another large hole on the inside of a hole that is situated about two feet from the back of the condyle.

The Os Linguale*, or Bone of the Tongue.—This is of a triangular shape, and its appendages make it appear an immense volume of bone; it is situated between the shoulders, and above the bones forming the sternum or bone of the chest or breast. The body of this lingual bone is curved in its form, the convexity of which projecting anteriorly, its inferior margin is crescent-shaped. There is to be observed a semilunar cavity at its smallest part, which, with the cartilages and ligaments, aids in the living animal towards forming the cavity of the throat. On its sides are some asperities, which give attachment to some of its powerful ligaments, &c. The top contains a deep sulcus or furrow, likewise lined with a similar surface, for the purpose of allowing origin and insertion to the muscles of deglutition. The bony appendages of this bone are two in number, and are articulated by means of loose ligaments to two extremities of the lingual bone. They are curved throughout their length, the convexity approaching inwards, where is a large and almost circular space in their upper part, that in the living animal contains enormous masses of fat. The remainder of these appendages are smooth, and appear only to give connection to a few muscular fibres.

THE STRUCTURE OF THE SPINE.

This portion of the skeleton is composed of fifty-four bones, and, with the head, forms a length which at first

---

* I use the term "*Os Linguale*," as more expressive of its use and situation, than the word "*Os Hyoides*" signifies particularly to the non-professional reader. See my "Synoptical Tables of an improved Nomenclature for the Sutures of the Cranium in Man, and the Mammàlia," &c.

sight seems impossible to have belonged to an animated
being, did we not know the creative power of an Almighty and wondrous God, whose works are

" By boundless love and perfect wisdom form'd."

As in most of the other mammàlia, we can divide the
bones of the spine into four series: cervical, dorsal, lumbar, and caudal.

Of the Vertebræ or Bones of the Neck.—The first
three of these have no spinal processes ; but it appears,
from its projecting from the cranium, that the spinal
marrow passes from the brain into its proper canal,
which is formed by the three lateral processes of the
first three bones of the neck.* This, it is to be observed,
is only a supposition of the superintendent of the dissection, Dr. Dubar, who states that the soft parts were
in such a state of decomposition, that it was almost impossible to distinguish it. The true spinal canal commences at the fifth cervical vertebra, and extends nearly
to the last caudal vertebra, being lost at the fifty-fourth.

This canal is formed of a triangular shape, by a series
of spinous processes which make the arch on the bodies
of the bones composing the spine, for the reception of
the spinal marrow.

The Atlas, or the first Bone of the Neck.—The anterior surface presents two articular fossæ (cavities) for
the reception of the condyloid processes or knobs of the
occipital bone, and is the means of the head articulating
upon the trunk; superiorly and laterally there are two
canals capable of receiving the human little finger, which
give passage to the vertebral vessels. The two trans-

---

* Formé par des apophyses latérales des trois premières cervicales.—
Dubar.

verse processes are tuberose and asperated, for the attachment of the adjacent muscles and lateral ligaments, permitting the head to perform the various motions intended for it by the great Author of nature. The posterior surface of this bone exhibits nothing beyond a few irregularities by which it is attached to the bone behind it. There is no hole or foramen in the transverse process of the atlas, for the passage of blood-vessels and nerves.

The second Bone of the Spine is of a curious yet regular shape, presenting an oval figure, the great diameter of which is transverse. This bone has no spinous, but has two enormous transverse processes; each of which possesses a very large foramen, which exceeds that of the occipital bone by twice its diameter. The third, fourth, and fifth cervical vertebræ have double transverse processes, so that they do not, as in the second, form a complete foramen or hole ; and the fifth exhibits the rudiments of a spinous process. The sixth forms a curve on its body, that, when united, has its convexity downwards.

The Bones of the Back.—These are fifteen in number, although the ribs are but fourteen on each side; the first of which, being bicipital or two-headed, is united to the first two dorsal vertebræ. The transverse processes have at their extremity an articulating surface for the union of the ribs—a phenomenon peculiar to these animals; and, consequently, the motions of the ribs must be somewhat limited. The bodies and processes of these vertebræ are very large, and in substance they are more dense and compact than the other vertebræ; which may be supposed to be thus made stronger

" By Him who never errs,"

in consequence of their having to support the whole
weight of the chest, with the heart, lungs, &c. &c.; together with the fins or swimming paws, and shoulder
bones. This portion of the spine is curved; the convexity is upwards: by this means the cavity of the chest
is greatly enlarged.

The Bones of the Loins are sixteen in number, and
bear considerable resemblance to the preceding, and are
without any articulating surfaces ; but it may be here
observed, that there exists not the slightest vestige of
any abdominal limbs : and they are found attached to
these bones, by means of muscles, two little bones,
forming the ossa pubis; and this forms the only vestige
of a pelvis. But I shall revert to this subject presently.

The caudal Vertebræ, or those approaching towards
the tail, are eighteen in number, and have bony appendages at their inferior surfaces, with the exception of
the eight nearest to the tail, where the appendages disappear. This portion of the spine tapers towards the
extremity; and, where it joins the tail, it exhibits a
slight curve, the convexity of which is placed inferiorly.

Of the Ribs.—These are fourteen on each side, and
form the walls of the chest. The structure of these
bones is dense, firm, and compact; which, with their
size and thickness, renders the animal capable of resist-
ing the most violent shocks : with the exception of the
first, which is almost vertical, the others take a more
posterior direction. The head of the first rib is double,
and articulated with the transverse processes of the
seventh and eighth vertebræ by means of tubercles received into the articulated fossæ or cavities of the vertebræ. The dimensions of this rib are very considerable,
and the sternal extremity of it is much larger than the

sternum itself. There is but one actual sternal rib on each side, which is fairly articulated with the sternum; whilst the others are, as in most other mammàlia, connected to that bone by a thick and powerful intervening cartilage, of which only the first five pair are real true ribs, and form any attachment to the sternum: the others are united to each other as the false ribs usually are, and the last three are not connected at all; consequently, we not improperly denominate them floating ribs. The fourth rib is the longest, and measures nine feet in length; the others gradually diminish as they approach the fourteenth, which is the smallest. The appearance of the chest as a whole will give the spectator a very good idea of the framework of a small sailing vessel; and it is impossible to form any accurate idea of the dimensions of this cavity, without making an examination of the interior: there only it is that a true conception can be formed.

The Sternum or Breast-bone.—This bone, when contrasted with the immense dimensions of the chest, is very small and spongy in its texture; and the layer of compact osseous tissue covering it is so thin as hardly to be perceived. The shape bears some rude resemblance to a cross, the apex or top of which is carried forward. This bone gives attachment to the first rib on each side, and is composed of three bones connected by cartilage. It has two plain surfaces, and exhibits nothing else worth mention.

The Bones of the Pelvis.—The pubic bones, which I have already cursorily mentioned, may not unaptly be considered as appendages of the spine. They are extremely small, and each has somewhat a triangular shape; but one of the angles is elongated upwards, and they bear altogether no small resemblance to the marsupial bones found in the kangaroo and other animals of New Holland, &c. They are found floating in the muscular walls of the abdomen; and the only connection they have with each other is by a very loose ligament. From their position, they, as far as we can perceive, can be of very little service to the animal, inasmuch as they neither possess size nor strength sufficient to protect the generative organs, or to guard, during the pregnant state, the fœtus within. However, there is not the least doubt but these bones must answer some important purpose in the animal economy, else the allwise Architect of the universe would never, in his wisdom, have constructed an organ insubservient to some useful function.

THE ANATOMY OF THE THORACIC EXTREMITIES, OR PECTORAL FINS.

The whale being deprived of clavicles, or collar bones, the pectoral fins are composed of the shoulder-blades, and what are, strictly speaking, the pectoral fins.

The Scapula or Shoulder-blade.—This is placed on part of the last cervical vertebra, and partly on the first dorsal, which it partly covers; it is a very large bone, of which the superior part is semicircular, and the inferior nearly quadrangular. The external surface is extremely smooth: there is no spinous process; but one, analogous to the acromion process of other animals, projects about 15 inches beyond the neck of the scapula. This must afford attachment to some of the muscles; the remaining muscles must form connections with the smooth surface, or with the superior margin of the bone.

The costal or internal surface has several strongly marked prominences and canals, which diverge towards the semicircular margin. These canals are evidently produced by the ribs during the fœtal or infantile state of the animal. The superior semicircular margin has several strongly marked asperities, where several very powerful muscles are inserted. The anterior margin, which is the shortest, is likewise the thickest at the inferior part of this, and at the anterior angle; the coracoid process [so called from its supposed resemblance to a crow's beak], and the one analogous to the acromion just described, both of them projecting anteriorly, are separated by a very deep canal, which, in the recent state, is filled up with a very fat cellular tissue. The glenoid or articular cavity of the shoulder-joint is found at the anterior margin of this bone: it is very flat, and there appears to be no attachment of the scapulo-humeral ligament, from which the animal enjoys motion at this part to a greater extent than the other Mammàlia; for the head of the brachium or arm-bone, which is enormous, can ultimately employ all its surfaces; in fact, it can describe full two-thirds of a sphere. Besides, the pectoral extremity not being controlled by a clavicle, its actions are less likely to be limited. This is the largest flat bone in the skeleton, next to those of the head; its structure is rather spongy, being only covered with a thin layer of dense substance. The fins contain bones analogous to the superior extremities in man, which I proceed to describe.

The Os Brachii, or Arm-bone, is short, but thick; the head is directed obliquely from outwards to inwards, where it articulates itself with the glenoid cavity of the shoulder-blade by means of a smooth and even articular cartilage; its cubital extremity is almost flat, and is articulated by simple ligaments to the radius and ulna. All the external surface of this bone is asperated, for the insertion of the muscles of the shoulder, and also of those which give motion to the fin.

Of the Bones of the Forearm.—The radius is flat, larger and thicker than the ulna, and offers no striking peculiarity, except the hardness of its compact tissue. Its articulation with the body is the same as with the ulna. *The Ulna, or Cubitus,* forms the inferior margin of the forearm; it is flat but curved through all its length; at the brachial extremity is a flattened tuberose process, which gives origin to strong tendons passing to the extremity of the fingers. The carpal extremity is united to the hand by a powerful tendinous substance: all the body of the bone has externally a dense compact tissue of ossific matter. *The interosseous Space,* or cavity, between the radius and ulna, is very narrow; it has a very thick membrane, not unlike a piece of leather.

The Carpus, or Wrist.—This is composed of six large bones; some in the form of a cube, others in that of a cylinder. They appear to have no articulated surfaces; but, on the contrary, are at very great distances from each other, and seem as though they were fixed in a thick tendinous substance, which envelopes them on all sides; so that, to preserve these bones in their natural state, it was impossible for Dr. Dubar to pay any attention to their particular shape.

The Metacarpus, or Hand, is composed of four long and thick bones, the two middle ones bearing a resemblance, but not in magnitude, to the thigh-bones of an ox; and, with the exception of the index, the three others are united to one and the same bone. They are slightly curved, and are of an equal thickness throughout.

The Fingers.—The fingers are four in number, and

the two smallest are the longest and strongest. The first, or the index, has four phalanges; the second, seven; the third, six; and the fourth, five; each having a space for a nail.

These bones or phalanges are independent of those forming the metacarpus. All of them are separated from each other by long tendinous ligaments, which are very flexible. Thus we find great strength within a small space in this limb, because it was there required: thus illustrating the beauty of Providence, in accommodating every part to the office it is designed to perform.

The B. Rórqual, the subject of the preceding remarks, was originally the property of the King of Holland, it having been taken during the period he held the sovereignty of Belgium. However, in consequence of the crown of that kingdom having been placed on the head of his Majesty Leopold I. by the Belgic nation, the proprietors, fearing that if they took the skeleton back to the continent it would be claimed by both monarchs, particularly as the King of Holland intended very properly to present it on its return to the University of Leyden*—to avoid this, I understand the proprietors have embarked with this stupendous skeleton for the United States, where they intend to exhibit it. However, in point of equity, I consider it to be the property of the Dutch monarch, he having only disposed of it conditionally to Mr. Kessels, who was to return it to him at the expiration of six years, which period terminates in the course of the present year.

---

* His Dutch Majesty should have recollected the old proverb, " a bird in the hand is worth two in the bush,"—and at once have presented this magnificent skeleton to the University.

ORDER II.—PREDENTATE CETACEA;

OR, THOSE WITH TEETH ONLY IN THE ANTERIOR PART OF THE UPPER JAW.

GENUS III.

MONODONS OR NARWHALES.*

THE animals of this genus have a single opening from the blow-holes, situated over the nape of the neck, and instead of teeth have generally a single tusk, or horn (sometimes two), proceeding from one side of the snout, a considerable length, more or less spirally twisted. Their head is not so long or thick in proportion as the Balænóptera, but terminates in an obtuse snout. Their mouth is small, and entirely destitute of teeth, or horny laminæ. The upper surface and sides of the body are variegated with spots of different forms, and a longitudinal ridge extends from the origin of the tail to a considerable distance along the back. There are no dorsal fins on the proper narwhales, which at all resemble the preceding genus. These animals are generally found in the Greenland and Icelandic seas.

There is some ambiguity respecting the species of this genus. Linnæus described but one species, and zoologists are not yet decided whether there should be more

---

* SYNONYMES.—*Narwalus*—La Cépède.

Narwhale and *Unicorn* } of the Whalers.

admitted. At all events we know that there are two varieties of these animals, which are sufficient to prove
that they should be deemed as distinct species, inasmuch
as they differ by fixed and permanent characters.   Accordingly Count La Cépède has so distinguished them.
The Rev. Dr. Fleming, in an excellent paper on one of
these species,* has followed his example, although his
specific characters differ from those of the French naturalist.

SPECIES I.

MONODON MONOCEROS, OR COMMON NARWHALE.

" In the creation of this curious animal, the Omnipotent Architect has
not only exhibited a still further striking proof of his own power, but has
likewise shown that for some wise purposes unknown to man that he
can produce an enemy equal in power, though not in bulk, to the stupendous creatures I have been describing."

THE Monodon Monoceros, or Common Narwhale, is an
animal possessing almost colossal strength, inasmuch
as it precipitates itself upon every thing giving it the
least offence, and furiously rushes against the most
trifling obstacle.   Its habitual sojourn is among the ice
and icebergs of the Arctic Seas.   Here, in the vast empire of eternal frost, where darkness reigns for so great
a portion of the year, this giant of the frozen ocean

---

* Memoirs of the Wernerian Society, voi. i. p. 131.

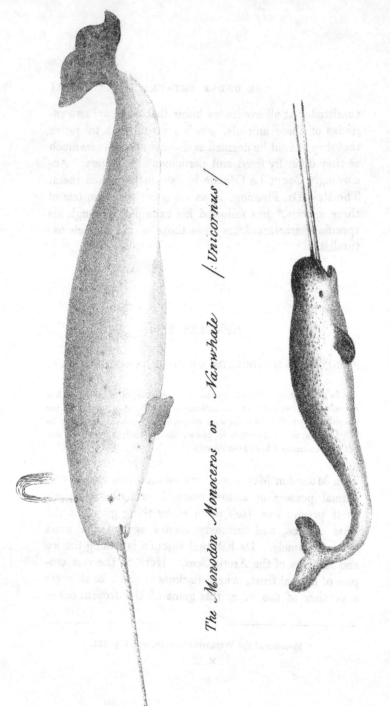

The Monodon Monoceros or Narwhale /: Unicornus :/

The Bicorned   Specimen discovered near Frieston Deeps, in Boston, Lincolnshire.

dares every power, braves every danger; and bent upon carnage, he attacks without provocation, combats without rivalry, destroying without necessity; and the only enemy to whom he is occasionally compelled to yield to, is man.

The form of the narwhale is ovoid, their usual length is from thirteen to sixteen feet in length, exclusive of the tusk, and in circumference (two feet behind the fins, where it is thickest,) from eight to nine feet. On perusing Baron Cuvier's work on the Animal Kingdom,* I was astonished at finding a great error into which this eminent naturalist had fallen, when he states the usual length to be from forty-two to sixty feet, whereas the longest I saw, and we caught several, was not more than fifteen feet.

Zoologists are indebted to Captain Scoresby, for the best description of the Narwhale. Previous to the publication of this gentleman's paper on this subject, very little was known respecting its peculiarities; I shall therefore quote his account, with a few remarks respecting its structure from my own observations and contemporary authors.

The form of the head, observes this gentleman, with the part of the body anterior to the fins, is paraboloidal, of the middle of the body nearly cylindrical, of the hinder part to within two or three feet of the tail somewhat conical, and from thence a ridge commencing both at the back and belly the section becomes first an

* I am much pleased to find that a cheap and elegant edition of this work is now publishing in English by Mr. Henderson, Old Bailey; from the Baron's last revised edition before his death; and in which I hope his former errors alluded to above will be corrected.

ellipse, and then a rhombus at the head to the junction
of the tail. At the distance of twelve or fourteen inches
from the tail, the perpendicular diameter is about twelve
inches, the transverse diameter about seven. The back
and belly ridges run half way across the tail or more;
and the edges of the tail in the same way run six or
eight inches along the body, forming ridges on the sides
of the rump. After a very slight elevation at the blow-
hole, the outline of the back forms a regular curve; the
belly rises, or seems drawn in, near the vent, and ex-
pands to a perceptible bump about two feet from the
genitalia. From the neck, three or four feet backward,
the back is rather depressed and appears flat.

The head is about one-seventh of the whole length of
the animal; it is small, blunt, and of a paraboloidal
form. The mouth is small, and not capable of much
extension. The under lip is wedge-shaped. The eyes
are small, the largest diameter being only an inch, and
are placed in a line with the opening of the mouth,
about thirteen inches from the snout. The spiracle, or
blow-hole, which is directly over the eyes, is a single
opening, of a semicircular form, about three inches and
a half in diameter or breadth, and an inch and a half radius or length. The fins, which are twelve or fourteen
inches long and six or eight broad, are placed at one-
fifth of the length of the animal from the snout. The
tail is from fifteen to twenty inches long, and three to
four feet broad. It has no dorsal fin; but in the place
of it, is an irregular sharpish fatty ridge, two inches in
height, extending two and a half feet along the back,
nearly midway between the snout and tail. The edge
of this ridge is generally rough, and the cuticle and rete
mucosum being partly wanting, appears as if it had
been worn off by rubbing against the ice.

The prevailing colour of the young narwhale is black-ish-gray on the back, variegated with numerous darker spots running one into another, and forming a dusky black surface. Paler and more open spots of gray on a white ground appear at the sides, disappearing altogether about the middle of the belly. In the elder animals, the ground is wholly white or yellowish white, with dark gray or blackish spots of different degrees of intensity. These spots are of a roundish or oblong form: on the back, where they seldom exceed two inches in diameter, they are the darkest and the most crowded together, yet having intervals of pure white among them. On the sides the spots are fainter, smaller, and more open. On the belly, they become extremely faint and few, and in considerable surfaces are not to be seen.

On the upper part of the neck, just behind the blow-hole, is often a close

Belly of the Narwhale.

patch of brownish-black without any white. The external part of the fins is also generally black at the edges, but grayish about the middle. The upper side of the tail is also blackish round the edges: but in the middle, gray, with curvilinear streaks upon a white ground, forming semicircular figures on each lobe. The under parts of the fins and tail are similar to the upper, only much paler coloured; the middle of the fins being white, and the tail of a pale gray. The colour of the sucklings is almost wholly a bluish gray, or slate colour.

The integuments are similar to those of the Balæ'na mysticetus, only thinner. The cuticle is about the thickness of paper: the rete mucosum three-eighths to three-tenths of an inch in thickness; the cutis or true skin thin, but strong and compact on the outer side.

A long prominent tusk, with which some narwhales are furnished, is considered as a horn by the whalers; and as such, has given occasion to the name of *Unicorn* being applied to this animal. This tusk occurs on the left side of the head, and is sometimes of the length of nine or ten feet: one caught by our seamen measured nine and a quarter, and according to Hans. Egede, they are generally fourteen or fifteen.* It springs from the lower part of the upper jaw, pointing forward and a little downward; being parallel in its direction to the roof of the mouth. It is spirally situated from right to left; is nearly straight, tapering to a round blunt point; is of a yellowish-white colour, and consists of a compact kind of ivory. It is usually hollow from the *Skull of the Narwhale.* bore to within a few inches of the point. A five feet tusk (about the average length) is about two and a quarter diameter at the base, one and three-quarters in the middle, and about three-eighths within an inch of the end. In such a tusk, there are five or six turns of the spiral, extending from the nose to within six or seven inches of the point. Beyond this, the end is without

---

* Description of Greenland, p. 77.

striæ, being smooth, clean, and white; the striated part
being usually gray and dirty. Besides this external
tusk, which is peculiar to the male, there is another on
the right side of the head, about nine inches long, firmly
imbedded in the skull. In females, as well as in young
males, in which the tooth does not appear externally,
the rudiments of three tusks will almost always be found
in the upper jaw. These are solid throughout, and are
placed back in the substance of the skull, about six
inches from its most prominent part. They are eight or
nine inches in length, both in the male and female; in
the former they are smooth, tapering, and terminate at
the root with an oblique truncation; in the latter they
have one extremely rough surface, finishing at the base
with a large irregular knob placed towards one side, which
gives the tusk the form almost of pocket pistols. Several
instances have occurred of male narwhales having been
taken which had two large external tusks. But this is
a very rare circumstance. Captain Scoresby states he
has never seen an external tusk on the right side of the
head; though he thinks it not improbable but that some
which have been shown him, having no perforation up
the centre, might be tusks of the right side. If I recollect
right, the late Mr. Brookes had in his splendid Zootomical
Museum, one or two specimens wherein a short right
tusk was seen. Sir Everard Home, in his examination
of the tusks of the narwhale, found, on sawing one, that
it appeared solid, in a longitudinal direction, " a hollow
tube in the middle through the greater part of its
length, the point, and the portion at its root, only being
solid,"* as is represented in the engraving on the next
page.

* Philos. Trans. for 1813, p.

With the exception of one, all the narwhales seen killed by Captain Scoresby, had a tusk from three to six feet in length, projecting from the left side of the head, of which about eight inches in length of each, was imbedded in the skull. The perforation, in all, extended from the base to within ten or twelve inches of the small end of the tooth.

The use of the tusk in the narwhale is not yet ascertained. It cannot be of use to them in procuring their food, for were this the case none would be exempt; neither is it perhaps necessary for their defence, otherwise the females and young would be subjected to the power of enemies without possessing the means of resistance, while the male would be in possession of an admirable weapon for its protection. The late lamented Dr. Barclay, Professor of Human and comparative Anatomy at Edinburgh, believed it to be principally a sexual distinction, similar to what occurs among some other animals.

Although it cannot be essential to their existence, it may however be occasionally employed. From the extremity being smooth and clean, while the rest is rough and dirty, and especially from the circumstance of a broken tusk being found with the angles of the fractured part rubbed down and rounded; it is not improbable but it may be used in piercing thin ice for the convenience of respiration, without being under the necessity of retreating into open water.*

* A splendid specimen of the tusk of the narwhale, measuring about ten feet in length, may be seen at the Widow Calvert's, worker in Ivory,

It cannot, however, be supposed, according to the opinions of Mr. O'Reilly and others, that it is used for the purpose of digging the sea plants from the rocks at great depths, nor with the intent of driving from their retreats the shrimps, molluscæ, vermes, and other minute animals which form his food, for this reason, that they are found commonly in deep seas, where the narwhale would be incapable of surviving under the immense pressure of the column of water resting on the bottom.

I may observe that the tooth or tusk of this animal (like the tusks of the *Trichecus Rosmarus,* or *Arctic Walrus*) is extremely compact in its tissue, and contains more phosphate and carbonate of lime than common bone. These animals do not appear to have any organs of voice; the females produce each a single young at birth, and this they nourish for several months with milk, supplied from teats that are situated near the origin of their tail. A quantity of fat or blubber, from two to three inches and a half in thickness, and amounting sometimes to about half a ton, encompasses the whole body. This affords a very large proportion of very fine oil.

The skull of the narwhale, like those of the delphinus, deductor, beluga, grampus, porpoise, dolphin, &c., is concave above, and sends forth a large flat wedge-shaped process in front, which affords sockets for the tusks. Upon this process is a bed of fat, extending to the thickness of ten or twelve inches horizontally (as the animal swims) or eight or nine perpendicularly. This fat gives a rounded form to the head; and, by its greater or lesser depositions, occasions a considerable difference in the

Tortoiseshell, &c., 189, Fleet Street, and another at a Comb Manufacturer's in Saint Martin's Court, Leicester Square. They are well worth the reader's examination.

shape and prominence of the forehead. In consequence
of this, what has been called the facial angle is in some
narwhales less than sixty degrees, in others upwards of
of ninety. The blow-hole communicates with a large
double cavity or air-vessel, immediately under the skin,
and this is connected with the naves of the skull, where
the opening is divided by a bony septum.

In a fine fatty substance about the internal ears of
the narwhale, are found multitudes of worms. They
are about an inch in length, some shorter, very slender,
tapering both ways, but are sharper at one end than
another. They are transparent. Within is the appearance of a canal; without is a brownish ridge, running
longitudinally along the body.

The spine of the narwhale is about twelve feet in
length, the cervical vertebræ seven in number, the
dorsal twelve, the lumbar and caudal thirty-five. The
whole are fifty-four, of which twelve enter the tail, and
extend to within an inch of its extremity. The spinal
marrow appears to run through the whole of the vertebræ from the head to the fortieth, but does not penetrate the forty-first. The spinous process diminishes in
length after the fifteenth lumbar vertebra, until it is
hardly perceivable at the nineteenth. Large anterior or
belly process on the opposite side of the spine and the
spinous process, attached to two adjoining vertebræ, commence between the thirtieth and thirty-first, terminating between the forty-second and forty-third vertebra.
The ribs are twelve on each, six true and six false, and
are slender for the size of the animal. The sternum or
breast-bone is of the shape of a heart, the broadest part
being forward. Two of the false ribs on each side are
united by cartilages to the sixth true rib; the rest are
detached.

The narwhale has no teeth, as the annexed cut of the lower jaw proves; its principal food consists of the molluscous animals, the smaller kinds of flat fish, and other marine animals. In the stomachs of several opened by Captain Scoresby, there were numerous remains of sepiæ or cuttle-fish.

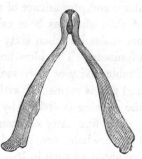

Lower Jaw of the Narwhale.

I have said that the females have no tusks, but Sir E. Home found two milk tusks in a skull brought on purpose to ascertain this object by Captain Scoresby, exactly resembling those of the male; they were eight inches long and imbedded in the skull, their points being only two inches and a quarter from the front of the skull. The only account on record appearing authentic of this circumstance, was by Anderson, who relates that Dick Peterson brought to Hamburgh, the skull of a female narwhale with two tusks, the left seven feet five inches long, and the right seven feet. Of such value were these tusks anciently considered, that even medical virtues have been attributed to them, and they have even been numbered among articles of regal magnificence. A throne made for the Danish monarch is said to be still preserved in the castle of Rosenberg, composed entirely of this substance, the material being considered as more valuable than gold.

The narwhale swims with considerable velocity; when respiring at the surface of the water, they frequently lay motionless for several minutes, with their backs and heads just appearing above the water. They are of a somewhat gregarious disposition, often appearing in numerous little herds of half a dozen or more together. Each herd is most frequently composed of animals of the same sex.

When harpooned, the narwhale dives with considerable velocity, and in the same way as the mysticètus, but not to the same extent. In general it descends about two hundred fathoms, then returning to the surface, is dispatched with the whale lance in a very few minutes.

The following table contains the dimensions of a male narwhale, killed near Spitzbergen, in 1817, from the observations of Mr. Scoresby:—

	Feet.	Inches.
Length, exclusive of the tusk	15	0
———— of the tusk, externally . . .	5	$0\frac{1}{2}$
———— diameter at the base	0	$2\frac{1}{4}$
———— from the snout to the eyes . .	1	$1\frac{1}{2}$
———————————— fins . .	3	$1\frac{1}{2}$
———————————— back ridge	6	0
———————————— vent . .	9	0
Circumference, $4\frac{1}{2}$ inches from the snout	,3	,5
———————— at the eyes and blow-hole	5	$3\frac{1}{2}$
———————— just before the fins . .	7	5
———————— at the fore part of the back ridge	8	5
———————————— vent	5	8
Blow-hole, length $1\frac{1}{2}$, in breadth . . .	0	$3\frac{1}{2}$
Tail, 14 inches	3	$1\frac{1}{2}$
Fins, 13 inches	0	$7\frac{1}{2}$

The heart weighed 11 lb.; the blood, an hour and a half after death, was at the temperature of 97.

Crantz* states that the tusk of the narwhale was

---

* Crantz's History of Greenland, 1767, vol. i. p. 112.

valued as a great curiosity, and sold excessively dear, until the Greenland fishery commenced, when the whalers found plenty of these animals in the northern part of Davis's Straits. He observes, they are so common in the north of Greenland that the natives, for want of wood, make rafters for their houses of them. The sailors, he continues, consider this animal as the harbinger or fore-runner of the common whale.

Dr. Shaw states that the narwhale may now be numbered among the *animalia rarioro* of the British zoology.*

SPECIES II.

MONODON MICROCEPHALUS, OR SMALL-HEADED NARWHALE.

THIS species* is of considerable size, varying from twelve to twenty-six feet in length, and of proportional circumference. Its body is of a narrower and more conical shape than that of the former species, and its head considerably smaller. Its upper surface is more flat and even. The colour of its upper parts is usually a dusky black, variegated with spots which, from being of a darker hue, are not very apparent. On the sides

* Shaw's Zoology, vol. ii. pt. ii. p. 476. Memoirs of the Wernerian Society, vol. i. p. 131.

the spots are more conspicuous, and of an oblong form. The belly is of a white colour. The whole skin is smooth and glossy, very thin, and closely united to the fat. The forehead is high, rising abruptly from the snout, proceeding afterwards for a few inches in nearly a horizontal direction, until it rises into a slight elevation, in the anterior part of which is situated the blow-hole. The mouth is extremely small, and the upper lip extends a little beyond the lower. The eye is situated almost below the orifice of the blow-hole, is about an inch in diameter, and has the iris of a chesnut colour. The individual of this species described by the Rev. Dr. Fleming " had one tooth [or horn] projecting from the left side of the upper jaw, and pointing a little downwards. There was no external appearance of any on the right side; and in the skull itself, only a small canal was observable, but no appearance of a horn. Its external length was twenty-seven inches, and the remaining portion of the base inserted in the socket twelve inches, thus making its whole length eighty-nine inches. The weight was twenty-eight ounces. In diameter at the base, where it entered the upper lip, it was one inch and a quarter. It was spirally grooved or twisted, and situated from right to left, and tapered from the external base to the point, which was blunt and solid. The part of the horn concealed in the skull was cylindrical and hollow." The horn of the narwhale found on shore at Frieston, a village near Boston in Lincolnshire, in 1800, was equal to one-third of the length of the animal :* in

---

* This was a bicorned specimen; it was twenty-five feet long, and the teeth six feet six inches, mouth rather small, teeth spirally twisted, terminating in a smooth point, as if worn down, and consist of very hard, compact ivory. Front of the head much rounded and blunt. Eyes black and small, considering the size of the animal. There was the rudiments

Dr. Fleming's specimen, it was little more than one-fifth. An account of the one just mentioned, found near Boston, was transmitted by the late Sir Joseph Banks, P.R.S., to Count La Cépède, though this gentleman has with the most unaccountable carelessness, described it as having been found in the seas which wash the coast of Boston, in latitude 40°, a description which evidently led to the supposition that the place was Boston in America. Dr. Fleming's specimen was found on shore at the entrance of the sound of Weesdale in Shetland, in 1808; it was only twelve feet long, and apparently had not attained its full size.

The narwhale is mentioned by the late Professor Walker of Edinburgh as having been frequently seen about the Shetland Isles, but he alludes to the common species, the *Monodon Monoceros*, the only species then known.*

---

of a fin on the back, and a hard ridge near the tail. Black above from the nose to the tail, softened with streaky spots towards the sides, which are white, with a few spots. Belly white, fins black. The whole animal was covered with a black and white horny substance, like some kinds of tortoiseshell, composed of laminæ for an inch or more in depth. In the stomach were found the horny beaks of cuttle-fish in great quantity. It was shown in Cockspur Street for some time, and also at Cambridge.— " *Sowerby's British Miscellany*," vol. i. p. 18.

I am indebted to the kindness of Mr. George Sowerby for the loan of the original plate from which my Lithographic representation is taken.

* Walker's Essays on Natural History and Rural Economy, p. 527.

GENUS IV.

THE ARNANACUS OR ARNANAK.

OTHO FABRICIUS, in his *Fauna Gröenlandica*, describes a species of cetàcea, called by the Greenlanders *Arnanak*, from the purgative qualities of its flesh and fat. The Abbé Bonnatérre, in his " Cetòlogie,"* considered it as a species of narwhale, and called it *Monodon Spurius;* but Count La Cépède, with more propriety, has formed it into a new genus, with the following character: —

" One or two small crooked teeth in the upper jaw; no teeth in the lower jaw, with a fin on the back."†

SPECIES.

ARNANACUS GRÖENLANDICUS,

OR

THE GREENLAND ARNANAK.‡

A SPECIES considered allied to the narwhale, but not perhaps, strictly speaking, belonging to the same genus; it has no teeth in the mouth, but from the extremity to the upper mandible project two minute, obtuse, conic teeth, a little curved at the tips, weak and not above an inch long: the body is elongated, black, and cylindrical. Besides the pectoral fins, it has a minute dorsal

---

* Encycloped. Méthodique. ART. Cetòlogie. p. 11.
† Hist. Nat. des Cétàcea, p. 164.
‡ SYNONYMES.—*Monodon Spurius.* Bonnatérre. Shaw. Gen. Zool. vol. ii. part 2. *Arnanak Gröenlandais.* La Cépède. p. 164.

fin. This species, it is to be remembered, is among the rarest of the cetàcea. The flesh and oil are considered as very aperient to those who partake of this delicacy. It inhabits the main ocean; it seldom approaches towards the shore, and feeds chiefly upon the *loligo* or *calamy*. It has a spiracle, and thus resembles the other species of the cetàcea. Both the flesh and oil are used by the natives of Greenland and Davis's Straits as food, but not without great apprehension of the qualities I have just mentioned. It is generally discovered dead, being but seldom captured alive.

The above is the description given by Fabricius,* which is the only account I have been able to find, but I have not been successful in procuring any accurate representation of this animal.

---

* Otho Fabricius, " Fauna Gröenlandica," p. 31.

AN ICEBERG IN LATITUDE 77° 42′ N.

ORDER III.—SUBDENTATE CETACEA,

OR, THOSE HAVING TEETH ONLY IN THE LOWER JAW.

GENUS V.

THE PHYSETERS,

THE CACHALOTS, OR SPERMACETI WHALES.

THE animals of this tribe are remarkable for the size of their head, which in some species is equal to the half and in others to a third of the whole animal. The upper jaw is excessively broad and deep, and has usually a few indistinct teeth almost covered with the gum; the lower jaw is long and narrow, enters into a fissure in the upper jaw, and is furnished on each side with a row of thick conical teeth, more or less obtuse. The blow-holes approach each other within the skull towards the anterior part of the snout, where they terminate in a common external opening. Below the snout is the principal cavity which contains the spermaceti, and it is chiefly from these animals that *ambergrease* is obtained. Some of them have a dorsal fin, others a callous protuberance on the back. There are several species. The oil obtained from these whales is denominated " *Sperma-Ceti Oil*," and is deemed to be superior to that derived from the preceding species.

SPECIES I.

PHYSETER MACROCEPHALUS,

THE GREAT-HEADED CACHALOT, or GREAT SPERMACETI WHALE.*

This species grows to the length of nearly sixty feet, is often thirty in circumference in the largest part of the head, the head forming by far the most conspicuous part of the animal, occupying the third part of the body. It has the appearance of an immense box, of a square form, angular at the sides, and truncated before. The upper is of much greater length than the lower, is also broader, its edge forming a very considerable projection, and folded back towards the centre, where there is an oval longitudinal cavity destined to receive the lower jaw. On each side of the upper jaw is a row of holes for receiving the teeth of the lower jaw, and in the interstices separating these cavities there are about twenty small teeth placed horizontally, just appearing a little above the jaw. It has been supposed that the teeth become longer, thicker, and more curved, in proportion to the age of the animal; the ordinary length being about six inches, and three inches in circumference at the base. These teeth are sharp on the side opposite to the place of insertion, but present a smooth, plain, and oblique surface, filling up the intervals that separate the cavities; which oblique surface is only visible, and from not attending to the form and disposition of their teeth, it has

---

* Synonymes.—*Parmacitty Whale,* in common language. *French Cachalot,* German. *Potfisch,* Dutch. *Potvisch,* Norwegian. *Kaskelot Potfisck Tweld-Hual,* Icelandic. *Grand-Cachalot,* Bonnat. Encyclo. *Cachalot Macrocephale,* La Cépède, p. 166.

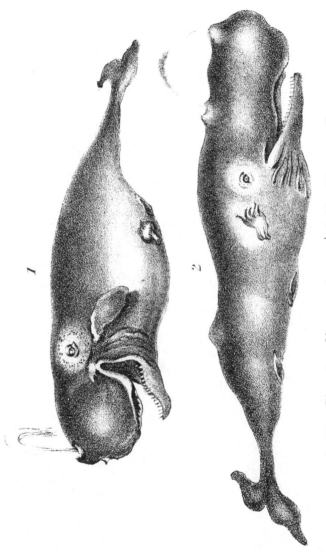

1 The Physeter Macrocephatus. | 2. The Physeter Trumpo.

Friedel. Lith. 24 Greek St.

been generally stated that the spermaceti whale had none in the upper jaw.

The tongue is a mass of flesh of a square form, of a livid red colour, filling almost the whole of the bottom of the mouth. The breathing-holes pass diagonally through the head, are about six inches in diameter, and unite into one just above the end of the snout. The eyes are black, small when compared to the bulk of the body, are furnished with eyelids, surrounded by strong short hair, though not very perceptible, and situated at a prodigious distance from the snout; the external orifices of the ears are scarcely discernible, the openings of which are placed behind the orbit of the eyes, on a cutaneous excrescence between the eyes and the pectoral fins. The swimming paws and tail are small for the size of the fish; the tail terminates in a fin, which is divided into two lobes: these are long and hollowed out in form of a sickle, and have a seamy margin.

The head is separated from the trunk by a transverse groove, extending to the place of insertion of the pectoral fins, which are of an oval form, three or four feet long, and three inches thick; the back is black, or of a slate blue, spotted with white, on which there is a callosity extending two-thirds of the whole length. It rises several inches above the surface, and is slightly inclined where it terminates, behind it is truncated. The belly is white, the flesh of a pale red like that of pork; the animal yields a considerable quantity of spermaceti, though this seems to vary in colour according to the climate in which the whale has lived. The fat or blubber is about five or six inches thick on the back, rather less on the belly; it lies immediately under the skin; it is not very productive of oil. This species feeds on lump, cuttle, and dog-fishes; it is said even to attack and

swallow the shark, one of which was vomited entire by
a spermaceti whale on being struck by a harpoon in the
Greenland Seas, where it is most commonly found. The
shark was four yards long, and on opening the stomach
of the whale, several fish's bones were found in it, and
some of a large size.

This whale inhabits chiefly the Greenland Seas and
Davis's Straits in North America ; has been occasionally
seen in the German Ocean, and on the European shores
to the southward. In 1784, a considerable number of
them were cast on shore on the coast of Lower Brittany
in France. The following are the dimensions of one of
them taken at the time :—

	Feet.	Ins.
Total length	144	6
From the anterior extremity of the snout to the eyes ..	8	0
From the eyes to the pectoral fins	3	0
From the pectoral fins to the organs of generation	19	7
Length of the tail..................................	6	9
Distance of the lobes of the tail	10	0
Circumference of the greatest thickness..............	34	8
Length of the upper jaw	5	0
—— —————— lower jaw..........................	4	6
Opening of the mouth	3	10
Breadth of the snout	5	0

The organs of generation resemble those of quadrupeds. The penis of the whale is enclosed in a sheath,
and is sometimes eight feet in length. On each side of
the genital organs in the female are placed the mammæ,
about four or five inches long. It is said to swim with
great velocity, and to rise very high above the surface of
the water. It is at this time that the fishermen take
the opportunity of striking him with their harpoons; and
it often happens that the parts of the body which have
been wounded become gangrenous and fall off before the
animal dies.

The bones are employed by the Greenlanders to form spears and other warlike weapons; the flesh, the skin, the fat, and the intestines, are applied to the same purpose as those of the narwhale. The tongue roasted is reckoned excellent food, and even esteemed a great delicacy. It is not very productive of oil.

Having described the natural history of this whale, I must not omit to mention the nature of the substance denominated "AMBERGREASE,"* which as a perfume is highly esteemed in the fashionable and commercial world. It is a solid, opaque, ash-coloured, fatty, inflammable mass, variegated like marble, extremely light, uneven, and rugged on its surface, and, when heated, emits a fragrant odour. It is found in the sea, or on the sea-coast, particularly in the Atlantic ocean, on the sea-coast of Brazil, and that of Madagascar; on the coasts of Africa, of the East Indies, China, Japan, and the Molucca islands; but most of the ambergrease brought to England comes from the Bahama islands, Providence Island, &c., where it is discovered on the coast. It is found in the abdomen of sperm whales, by the whalers, always in lumps of various shapes and sizes, from half an ounce to a hundred and more pounds.

There have been numerous opinions concerning the origin of this substance. But the most scientific and satisfactory account is that given by the late Dr. Swediaur.†

We are informed by all writers on this subject that sometimes claws, beaks, and feathers of birds, parts of vegetables, bones of fish, fish, shells, &c., are either found in it, or mixed with it. Of a very large number of pieces which the doctor examined, he found none that contained any such thing, although he allows that they may be sometimes found; but, in all the pieces of any considerable size found either on the sea,

* SYNONYMES.—*Ambergrise*, or *Grey-amber*. *Amber. gris.* FR.
† Vide Philosophical Transactions for 1783.

or in the whale, he always found a considerable number of black spots, which, after the most careful examination, appeared to be the beaks of the *Sepia octopedia*.* And these, he thinks, might be the substances mistaken for claws and beaks of birds, or pieces of shells.

According to the best information Dr. Swediaur obtained from several of the most intelligent persons employed in the spermaceti whale fishery, and in procuring and selling ambergrease, it appears that this substance is found in the abdomen of the whale, but only in the species I have been describing. The New-England fishermen, according to their accounts, have long been aware that this substance was found in the spermaceti whale, and were so convinced of this fact, that whenever they hear of a place where ambergrease is to be met with, they directly conclude that the sea, in that part, is frequented by the *Physeter Macrocephalus*. The sperm whalers, whenever they strike a spermaceti whale, observe that it constantly rejects whatever it has in its stomach, and also discharges its fæces at the same time; if this latter circumstance occurs, they are generally disappointed in finding ambergrease in its belly. But whenever they discover a spermaceti whale (either male or female) which seems torpid or sickly, they are always pretty sure of finding ambergrease, as the whale in that state seldom voids *per anum* on being struck. They likewise generally meet with it, in the dead spermaceti whales, which are sometimes found floating on the surface of the sea. It is observed, likewise, that the whale in which they find ambergrease often has a morbid protuberance, or, as they express it, a species of gathering, in the lower part of its abdomen, in which, if cut open, the ambergrease is found. It has also been observed, that all those animals possessing this substance seem not only torpid and sick, but are constantly leaner than others ; so that, if we may judge from the constant union of these two circumstances, it would seem

---

* The cuttle-fish.

that a larger collection of ambergrease in the abdomen of the whale is a source of disease, and probably the cause of its death. As soon as the whale seamen capture a whale of this description, torpid, sickly, emaciated, or one which retains its fæces on being struck, they immediately either cut up the before-mentioned protuberance (if there be any), or they rip open its bowels from the anal orifice, and find the ambergrease sometimes in one, at others in different lumps, varying from one pound to twenty or thirty pounds in weight, and from three to twelve inches in diameter; they are generally found at the distance of two, but most frequently at about six or seven feet from the anus, and never higher up in the intestinal canal; which, according to their description, is in all probability the *intestinum cœcum*,* hitherto mistaken for a peculiar bag made by nature for the secretion and collection of this singular substance. That the part they cut open to get the ambergrease is no other than the intestinal canal is certain, because they commence their incision at the anus, and find the fæces, which it is impossible to mistake. The ambergrease they find is not so hard as that obtained on the sea or sea-coast, but soon grows hard in the air; when first taken out it has nearly the same colour, and the same disagreeable odour, though not so strong as the liquid fæces; but, on exposing it to the air, it by degrees not only grows greyish, and its surface is covered with a greyish dust, like old chocolate, but it also loses its disagreeable smell, and, when kept for a certain length of time, acquires the smell which is so agreeable to most persons, and particularly by the fair sex, with whom its essence forms an essential ingredient of the toilet.

In considering whether there is any material difference between the intestinal and the sea or sea-coast ambergrease, Dr. Swediaur refutes the erroneous opinion, that all ambergrease found in whales is of an inferior quality, and therefore

---

* The first of the large intestines.

much less in price. Ambergrease, he observes, is only valued for its purity, lightness, colour, compactness, and smell. There are pieces found on different coasts, which are of a very inferior quality; whereas, there are often found in whales pieces of it of the first value; nay, several are found in the same whale, according to the above-mentioned qualities, varying in value. This substance hardens in the air, and improves by age. It is more frequently discovered in males than in females; the pieces in the latter being generally smaller; whereas those in the males seem constantly to be larger and of a better quality; and therefore the high price in proportion to the size is not merely imaginary for the rarity of it, but is in some respects well founded, because such large pieces appear to be of greater age, and possess all the qualities which I have mentioned, in a higher degree of perfection than the smaller pieces. As the *Sepia octopedia* (or cuttle-fish) forms a part of the natural food of the *Physeter Macrocephalus*, the whalers know from experience that whenever they find any recent remains of it swimming on the surface of the ocean, they conclude that a whale of this species is, or has been in the neighbourhood. Another circumstance which corroborates the fact is, that the spermaceti whale on being struck generally rejects by the mouth some remains of the sepiæ. Hence it is easy to account for the pieces of these animals which are found in ambergrease. The beak of the sepia is a black horny substance, and therefore passes undigested through the stomach into the intestinal canal, where it is mixed with the contents of the alimentary tube; after which it is either evacuated with them, or, if these be preternaturally retained, forms concretions with them, rendering the animal sick and torpid, producing an obstipation, which ends either in an abscess, as has been frequently observed, or terminates the life of the animal; whence, in both cases, on the bursting of the abdomen, that hardened substance, known under the name of *ambergrease*, is found swimming on the sea, or thrown upon the coast.

From the preceding observations, Dr. Swediaur concludes that all ambergrease is generated within the bowels of the *Physeter Macrocephalus*, or spermaceti whale; and there mixed with the beaks of the *Sepia Octopedia*. He therefore defines *ambergrease* to be the preternaturally hardened fæces of this species of whale. Mr. Magellan has mentioned the existence of a vegetable *ambergrease*, gathered from a tree which grows in Guyana, and is called *cuma*. Specimens of this were presented to him by Mons. Aublét, author of the "Histoire de la Guyane," published in 1774, who himself collected it on the spot; and Mr. Magellan presented a portion of it to the late Dr. Fothergill of Bath, and the late Dr. Combe. It is of a whitish-brown colour, with a yellowish shade; melts and burns like wax on the fire, but is rather of a more powdery consistence than any ambergrease he had seen.

The use of ambergrease in Europe is now nearly confined to perfumery, though it has been recommended in medicine by several eminent physicians. It is soluble in boiling spirits of wine; from which, if the saturated solution be set in a very cold place, a part of the ambergrease concretes into a whitish unctuous substance. Distilled, it yields an aqueous fluid, a brown acidulous spirit, a deep-coloured oil, a thicker balsam, and sometimes a concrete salt. The spirit, oil, balsam, and salt, are similar to those obtained from amber, except that the oil is more agreeable to the smell. Rectified spirit of wine takes up near one-twelfth of its weight of ambergrease. According to Neumann, if the spirit is impregnated with a little essential oil, the ambergrease will dissolve more readily in it. A deep-coloured tincture is made with alcohol, but not a stronger. Acids and alkalies have no effect upon it; water and expressed oils have as little. It is one of the most agreeable perfumes; it heightens the natural odour of other bodies; but the great secret to this end is to add it sparingly, that while it improves the smell of that to which it is added, its own may not be discovered. In the dose, from two grains to a scruple, it is a high cordial, and powerful antispasmodic; although the common dose is from two grains to four, which may be given in an egg lightly poached. Riverius states that ambergrease is a specific against the *Rabies Canina*,* or hydrophobia.

In Asia, and part of Africa, ambergrease is not only used in medicine, and as a perfume, but considerable use is also made of it in cookery, by adding it to several dishes as a spice. A great quantity of it is constantly bought by the pilgrims who travel to Mecca, who probably offer it there for the purpose of incense; in the same way that frankincense is used by the clergy in the performance of the sacred ceremonies of the Roman Catholic church.

Ambergrease may be known to be genuine by its fragrancy, which it emits when a hot needle or pin is thrust into it, and its melting like fat of a uniform consistence; whereas the counterfeit will neither yield such a smell, nor fatty texture. There is another substance afforded by this species of the physeter, of great importance in a commercial point of view, it is denominated

SPERMACETI.†

This is a substance of a whitish, flaky, and unctuous nature, prepared from the oil, which is principally derived from the blubber of this species. Although known to the ancients, they were unacquainted with the nature of its preparation; and even Schroëder doubts whether he should consider it as an animal or mineral substance. The spongy, oily mass from which it is made is found in a large triangular trunk, four or five feet deep, and ten or twelve long, filling almost the whole cavity of the head, and seeming to be entirely different

---

* It would be perhaps worth while to give it a fair trial in this terrific of all diseases, particularly as every other known remedy has failed.

† In vulgar language it is called *Parmasitty*. It derived its name, Spermaceti (seed, or sperm of the whale), no doubt, from an avaricious desire to raise its commercial value, or from a notion of its scarcity. *Rees's Cyclopedia*.

from the proper brain of the animal. The oil is separated from it by dripping through a filter. In this state it has a yellow, unctuous appearance, and is brought to England in barrels. An ordinary sized whale will yield upwards of twelve large barrels of crude spermaceti. It is purified as follows : the mass is put into hair or woollen bags, and pressed between plates of iron in a screw-press until it becomes hard and brittle. It is then broken into pieces and thrown into boiling water, where it melts, and the impurities rising to the surface, or sinking to the bottom, are separated from it : on being cooled, and removed from the water, it is put into fresh water in a large boiler, and a weak solution of potash (of commerce) gradually added. This is thrice repeated, after which the whole is poured into coolers, when the spermaceti concretes into a white semitransparent mass, and, on being cut into small pieces, assumes the flaky appearance which it has in the shops.

The great use of spermaceti is for making candles, and it is also employed in medicine, but its healing virtues are chiefly imaginary, being not so beneficial as the more bland vegetable oils. Spermaceti candles are of modern manufacture ; they are made smooth, with a fine gloss, free from rings and scars, and superior to the finest wax candles * in colour and lustre, and, when genuine, leave no spot or stain on the finest silk, cloth, or linen. It melts at 112° Fahrenheit, and a higher temperature evaporates, with little alteration. Its specific gravity is 9.433, and, by the aid of a cotton wick, it burns with a clear white flame, superior to that of tallow, and without any disagreeable odour.†

* With the exception of the " Diamond wax candles," recently invented by Dr. Bulkeley, which surpass any I have hitherto seen in beauty of appearance and brilliancy of flame ; they are well worth the patronage of my readers.

† For the chemical analysis of spermaceti, or, as M. Chevreul more properly calls it, *Cetine*, the scientific reader is referred to Dr. Ure's Dictionary of Chemistry, *article* FAT.

SPECIES II.

PHYSETER CETADON, OR THE SMALL SPERMACETI WHALE.*

By taking the teeth as the most certain characteristics of different species, according to the opinion of most naturalists, we shall easily distinguish this species from the others. In this species the head is of a round form; the opening of the mouth is of a moderate size; the lower jaw is longer, but not so broad as the upper. It is furnished with a row of teeth on each side; and these correspond to the cavities in the upper jaw which receive them. Here we shall find a peculiar structure of the teeth in this species; for the teeth are curved and blunt, and that part of the tooth which rises above the gum has a greater thickness than where it is inserted into the jaw; each tooth is flat at the top, and marked with con-centric lines. The longest are two inches in length, and about an inch in circumference at its greatest thickness.

Sibbald has mistaken the breathing-hole for nostrils, which mistake seems to have arisen from the breathing-hole being near the snout of the fish.

This species chiefly inhabits the northern seas, though, towards the end of the seventeenth century, 102 of this species came on shore at Cairston in the Orkney Islands. The longest was twenty-four feet.

---

* Synonymes.—French *le petit Cachalot*; Norwegian, *Swine-Hual*; Greenland, *Kagutilik.*

SPECIES III.

PHYSETER TRUMPO, OR BLUNT-HEADED CACHALOT *

THE length of this species is nearly sixty feet, and its
breadth about fifteen. The head is of an immense size,
dividing the body nearly into two equal parts. The
upper jaw is at least five feet longer than the lower;
round and obtuse at the snout, and about eight feet
deep (from crown to base). The lower jaw is about ten
feet long, very narrow, is furnished with eighteen teeth,
straight and all pointing outwards, about three inches
distant from each other; and when the mouth is shut are
received into cavities of the upper jaw. The eye is
many feet behind the snout, nearly in the middle of the
breadth of the upper jaw, is small, over which is a considerable convexity and a similar prominence, below the
articulation of the swimming-paws, which are proportionally larger than in the former species. The breath-
ing hole is at least a foot in diameter, and is placed at
the superior extremity of the snout.

The body is irregularly conical, with a prominence on
the back, in place of the dorsal fin, which is more than
a foot thick; its thickest part is near the insertion of the
pectoral fins. These are very small, and that of the tail
is divided into two lobes, measuring fifteen feet from tip
to tip, and another prominence on the belly just before
the anus, the penis is eight feet long. The colour of
the skin is a blackish grey, very soft to the touch. It is

---

* SYNONYMES.—French, *Le Cachalot de la nouvelle Augleterre*;
Le Trumpo. Physeter Macrocephalus. Var. [Linn. Syst. Nat. Dudleii]
Baleena, Klein. Blunt-Headed Cachalot. Penn. Brit. Zoolog. Vol. iii.
Cachalot Trumpo. La Cépède, p. 112.

chiefly an inhabitant of the seas which wash the shores of New England. It is found in the Greenland Seas; and is occasionally seen off the coast of France and Britain. In 1741, one was taken near Bayonne in France; it yielded ten tons of spermaceti, which was reckoned of a superior quantity to that of the large spermaceti whale. In its stomach was found a round mass of seven pounds weight, which was taken for Ambergrease.

The substance called spermaceti is lodged in particular cells in the head, near the seat of the brain. It is extracted by making a hole in the skull.* This species yields a very fine and prodigious quantity of it; and its blubber is very productive of oil of a finer quality than that of the Greenland whale.

The blunt-headed Cachalot is a bold and daring animal, and swims with great swiftness, and is observed by some naturalists to be more dangerous than any other of the same species, and when attacked turns on its assailants with open mouth, with which it defends itself, and throws itself on its back.

If we attend to the form of the body, the structure of the head, the number and structure of the teeth, it seems to constitute a distinct species. The Spermaceti whale thrown ashore near Bayonne was described to be of the following dimensions :—

	Feet.	In.
Total length	49	0
Greatest circumference at the Fins	27	0
From the extremity of the tail fin, to the opening of the anus	14	0
Length of the penis	4	0
Sheath inclosing it	1	6
Diameter of the penis	0	7
Distance of the extremities of the two lobes of the tail	13	6

* See pages 156 and 157.

SPECIES IV.

THE PHYSETER CYLINDRICUS,

OR

THE ROUND SPERMACETI WHALE.

THIS species has a hunch on the back; the teeth are curved and pointed at the top, the spiracles or breathing-holes are in the middle of the snout.

The form and relative situation of the trunk and head, the position of the spiracles, the relative length of the jaws, the number and structure of the teeth, and especially the size of the dorsal fin, present differences sufficiently distinguishing this from the following species. The body is cylindrical from the extremity of the snout to a line drawn perpendicular to the place where the penis is inserted, and from thence to the tail fin it gradually diminishes. The head is at least one-third of the whole length of the body. The profile of the head presents a kind of parallelogram. The jaws are nearly of equal length. On each side of the lower jaw there is a row of twenty-five curved, sharp-pointed teeth. The breathing-holes are placed in the centre of the superior extremity of the snout. The dorsal fin is replaced by a hunch, eighteen inches high, and four inches and a half long at the base. The tail fin is divided into two lobes, forming a kind of crescent. One of this species is described by Anderson forty-eight feet long, twelve of perpendicular height, and thirty-six in circumference at its greatest thickness.

The colour of this whale is uniformly blackish. It inhabits the Arctic Seas, and the northern parts of the North Atlantic Ocean.

M

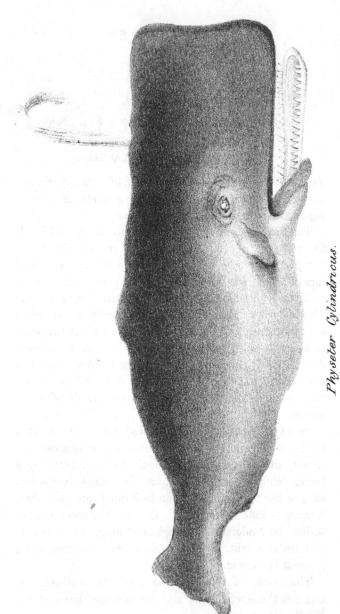

Physeter Cylindricus.

SPECIES V.

THE PHYSETER MICROPS,

OR

THE SMALL-EYED CACHALOT.

THERE is considerable confusion in the accounts of naturalists who have treated of this species of whale; and this probably arises from their not having attended sufficiently to the form of the teeth. According to the words of La Cépède, " the Physeter Microps is one of the largest, most cruel, and most dangerous inhabitants of the deep. Adding to formidable weapons, the two great sources of strength, bulk and velocity, greedy of carnage, a daring enemy, and an intrepid fighter, what part of the ocean does he not stain with blood ?"

Its head is so enormous as to equal the whole length of the animal, independent of the tail fin, and it is as large in circumference as any part of the body. The upper jaw, though not extending quite so far as the snout, properly so called, is a little longer than the lower jaw. The teeth which appear in this latter are conical, curved, hollow towards the roots, and set into the jaw about two-thirds of their length. The part beyond the gum is white like ivory, and its tip acute, and curved first towards the throat, and then a little outwards. According to Fabricius, there are only twenty-two teeth in

---

* French, *Cachalot Microps, Cachalot à dents enfancille*. Norweg. *Staur-Hyming ;* Greenland, *Tisagusik*. The *Parmacetty Whale*, or *Poteval-fish*, Dale's Harwich, *Black-Headed Spermaceti Whale* of Commerce. *Cachalot Catodon* or *Pot-fish*. Crantz's Greenland, *Great-Headed Cachalot*. Pennant Brit. Zool. vol. iii. *Cachalot Microps*. Bonnatérre, *Physetere Microps*. La Cépède, p. 227.

the lower jaw, eleven on each side; but according to the most respectable naturalists the teeth are forty-two in number. The upper jaws have cavities for receiving the teeth of the lower; and between them there appears to be short blunt teeth, almost entirely hidden by the gums. Each tooth extends to a finger in length, and is about one inch and a half broad, the longest occupying the middle part of the jaw, the smaller being at the extremities. The snout ends in a blunt surface, and according to most naturalists the lower jaw is the longest. The Greenlanders say that this whale has teeth in the upper jaw, but this is not clearly ascertained.

The swimming-paws are about four feet long. The dorsal fin is straight, high, and pointed, and by some zoologists has been compared to a needle. The whole length of the animal usually exceeds fifty feet, and the skin is of a black colour. It is found in the Arctic Ocean, and has occasionally made its appearauce in the North Sea. Sir Robert Sibbald mentions one, which was cast ashore on the coast of Scotland; and so lately as 1769, one of this species was stranded at Cramond, a little above Leith, in the Firth of Forth, and attracted many thousands of spectators from Edinburgh and the surrounding country. Count La Cépède relates an account of seventeen which appeared in 1723, in the mouth of the Elbe, and were mistaken by the fishermen of Cuxhaven for so many Dutch fishing-boats.

This animal attacks not only porpoises and other small cetàcea, but likewise the largest species of the balænópteræ, especially the B. Acuto-Rostrata, on which it fastens with its crooked teeth, and tears immense pieces from their bodies. It is also said to pursue the young Greenland whale, which it compels to fly for refuge through the boundless ocean.

Its flesh is esteemed as a great delicacy by the Esquimaux, and the natives of East Greenland and Davis's
Straits; it yields a quantity of spermaceti, but very little
oil.

The Count La Cépède supposes this animal to have
been the sea monster, from which Perseus delivered the
fair Andromeda; and he labours to prove that the *Orca*
described by Pliny as having been attacked in the port
of Ostea, by the Emperor Claudian at the head of his
troops, was not a *Grampus,* as is commonly supposed,
but on the contrary a *Physeter Microps.*

SPECIES VI.

THE PHYSETER MULAR,

OR

THE GREAT FINNED CACHALOT.*

This species is distinguished by a very elevated fin,
situated on the middle of the back. The teeth are
slightly curved and obtuse.

According to Mons. Brisson, this Cachalot resembles
the preceding in the formation and general structure of its
body. It however differs in the form of the teeth, which
are less curved, and are obtuse. The longest, which are
eight inches in length, and nine inches in circumference,
occupy the front of the jaw. The others are only six
inches long. Sometimes the teeth are found to be hollow, and sometimes they are solid. A question here
arises, which like many others in Natural History is
difficult of solution. Is this anatomical difference owing
to the *age of the individuals in which it is observed?*
Beside the pectoral fins, that which is placed on the back

---

* Synonymes.—*Le Cachalot Mular.* Bonnatérre. *Great Finned
Cachalot.* Dewhurst.

is very remarkable on account of its length.   Sir Robert Sibbald compares it to the *mizen mast* of a vessel.

According to Captain Anderson, this species is farther distinguished by having three bunches or protuberances towards the extremity of the back; the first is eighteen inches high; the second six inches; and the last only three inches.   The same historian observes, that he was informed by the captain of a whale ship, that he saw on the coast of Greenland a great number of this species of whale, at the head of which was one of *one hundred* feet long, and which appeared to be the leader, which, on the appearance of the ship, gave such a terrible shout, spouting water at the same time, as to shake the vessel. At this signal the whole made a precipitate retreat.

This species of the Cachalots is gregarious, frequenting the seas about the North Cape, and the coasts of Finmark, the Atlantic Ocean, the Arctic Icy Ocean, and particularly in the Greenland Seas, and about the Orkney Islands.*   They are very wild and difficult to wound, and are consequently but rarely taken.   It appears that the harpoon can only pierce them in one or two places near the pectoral fins.

The fat or blubber is very tendinous, and yields but a small proportion of oil.

SPECIES VII.

THE PHYSETER BIDENS SOWERBY*I*,
OR THE
TWO-TOOTHED CACHALOT OF SOWERBY.

MR. SOWERBY, to whose kindness I am indebted for the representation and description of this species of

---

* Histoire Naturelle des Cétacées, p. 242.

whale, and which was published in the first volume of
the British Miscellany by his father in 1806, considered this animal as a new species of the Cachalot
family, and through Mr. James Brodie, the celebrated
naturalist (and late member of parliament for Forres),
he was enabled to add it to the present list of British
zoological subjects, it having been first observed by
this gentleman, from the circumstance of its having been
cast on his own estate, near Brodie-house, Elginshire.
On account of its great weight and bulk, he was only
able to send Mr. James Sowerby the head; which to
such an eminent zoologist was sufficient to distinguish
it from all others of the same genus, and to form a
beautiful specimen for Mr. Sowerby's Museum.

On examination of the mouth, this gentleman was
both pleased and astonished to find, from the extraordinary formation of its mouth, and the situation of its
teeth, that it was likely to prove a species not yet described, and he was soon confirmed in this opinion when
it was exhibited to some of the most eminent zoologists,
at one of the *converzationes* of the late Sir Joseph
Banks, in Soho Square.

From the description of this creature, according to
Mr. Brodie and Mr. Sowerby, it appears that the cuticle
on every part of the head and body was perfectly pellucid
and satiny, reflecting the sun to a great distance. Immediately beneath the cuticle, the sides were completely
covered with white vermicular streaks, in every direction,
which at a little distance appeared like irregular cuts with
a small sharp instrument. It was of the male sex.

The animal was oblong, black above, nearly white
below, sixteen feet long, eleven feet in circumference at
the thickest part, with one fin on the back. The head
acuminated. Lower jaw blunt, longer than the upper,

with two short lateral *bony teeth,* which forms the principal distinguishing characteristics of this species. The upper jaw was sharp, let into the *lower one* by two lateral impressions corresponding with the teeth. The opening of the mouth measured one foot six inches. Tongue smooth, vascular, and small. The throat rough and also vascular; beneath it were two diverging furrows, terminating below the eyes; which were small, and placed six inches behind the mouth. Mr. Sowerby observes, " We know of no whale with only two teeth in the lower jaw, described by any author. Gmelin mentions one with two teeth in the UPPER JAW, which he calls *Balæ'na Rostrata.* Johnson has figured what he calls *Delphinus Fæmina* with apparently two teeth in the *upper jaw,* and impressions in the lower one.* We cannot be mistaken as to the position of the head in our figure, for the spiracle (which was lunated, the ends pointing anteriorly) was sufficiently conspicuous when it was reclined. We might have called it *Physeter Rostratus,* with some propriety; but this might have created confusion. It is however a curious circumstance, that such an appellation would suit better if it were described with the wrong side upwards; which will be easily observed, if the plate be reversed: and the jaws, in this case, very aptly resemble a bird's beak."†

As Mr. Sowerby has been the first naturalist that described this whale, I propose calling it " *The Physeter Bidens Sowerbyi.*"

---

* These appear to be the same as Schreber's figure, which is marked "*Delphinus bidens;*" it is not unlike our animal, but if meant for the same it is represented rather too short, with the head the wrong side upwards.

† Sowerby's British Miscellany, p. 2.

SPECIES VIII.

THE PHYSETER GIBBOSA,

OR

THE BUNCHED CACHALOT.

THE *Physeter Gibbosa** is a native of the Northern Seas,
and is said to have the same general form as the *B. Mysticète,* with the exception of its being of much smaller
dimensions, and having the back furnished with one or
more tubercles, which have been denominated *Bunches,*
and hence gave origin to its common appellation. A
variety of this species is found on the coasts of New
England ; and another, having six tubercles along the
back, was supposed by the late Dr. Shaw to inhabit the
coasts of Greenland, but neither of these varieties appears to be very accurately known either by the Whalers
or Zoologists. Their baleen is said to assume a pale or
whitish colour, and in their lower jaw they resemble the
Cachalots, by having teeth. The annexed plate is an
accurate representation of this animal.

---

* SYNONYMES.—*Balæ'na Mocra* or *lean* whale, Dr. Klein. In German, *Knotenfisch.* In Dutch, *Knobbefisch.* The Bunched Whale of
the English zoologists.

ORDER IV.—AMBIDENTATE CETA'CEA.

OR THOSE HAVING TEETH IN BOTH JAWS.

GENUS VI.

DELPHINUS OR DOLPHINS.

THE animals of this genus are smaller than most of the preceding tribes; the largest species hardly exceeds twenty-five feet. Their jaws are lengthened considerably, but are of equal length, and are each furnished with a row of conical teeth, more or less numerous in the particular species. The spiracles or blow-holes, after traversing the upper jaw, unite without in a single orifice, which is in the form of a crescent, and is situated at the top of the head. Their eyes are placed near the angle of the mouth. All the species but one are furnished with a dorsal fin, which is sometimes of an enormous length. They are not confined to the Arctic Ocean, but in fact are found in most seas, and in the mouths of most rivers.

The exact number of species is not accurately ascertained, but the following are most commonly known to zoologists.

SPECIES I.

THE DELPHINUS PHOCEANA COMMUNIS,

OR

THE COMMON PORPOISE.

THE Porpoise* may justly be denominated as the most common of the *Order Cetàcea,* inasmuch as it is equally common to the Baltic Seas, the coasts of Greenland and Labrador, in the Gulf of Saint Lawrence, throughout the whole of the Atlantic Ocean, in the great Pacific Ocean, in the Gulf of Panoma, Mexico, and California, near the Gallapagos Isles, and in short in almost every sea.

The porpoises are constantly to be found sporting in the stormy ocean, traversing its agitated surface in the most complete tranquillity. They cut without difficulty the foaming wave, and when the black tempest appears to convulse the sea, even to its profoundest abysses, they float upon its bosom with the same security as in the sunny hour of perfect calm. It is to its muscular force, and the powerful instruments of notation with which nature has provided it, that the porpoise owes the astonishing rapidity of its motions, and those wonderful springs and evolutions which the eye can scarcely follow. This animal exhibits the figure of a very elongated cone in its body and tail. The head may be considered as

---

* SYNONYMES.—*Tursio,* Pliny and Rondélet, Hist. des Poissons. *Meerschwein oder tunin,* Martin. *Le Marsoin,* Belon, Hist. des Poissons. *Delphinus Corpore fere coniformo, dorso lato rostra-sub-acuto,* Artédi Synop. The *Porpesse,* Willoughby, Ray, Pennant, and Shaw. *Niser ou le Marsoin,* Egede. *Dauphin Marsouen,* La Cépède, Hist. Nat. de Cetacées, p. 287.

1. Delphinus Phoceana or Porpoise. | 2. Delphinus Didelphis or Dolphin

Friedel Lith. 24 Greek St

another very short cone, the base of which is closely
united to that of the former; it is slightly inflated above
the eyes, which are small and situated at the same elevation as the division of the lips; the iris is yellow and
the pupil triangular. Beyond the organ of vision, and at
a small distance, is that of hearing, but which from its
diminutiveness is difficult to distinguish.

The tongue is broad, soft, and flat, and apparently
indented at its edges.

The orifice of the two spiracles is in the form of a
crescent, whose convexity faces the tail. The respiratory organ is placed above the space comprised between
the eye and the aperture of the mouth.

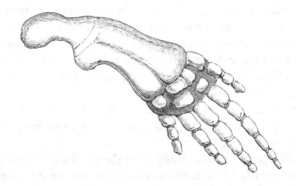

The pectoral fin, the bones of which are represented
in the above engraving, bears a great resemblance to
the human hand; it is situated very low, and almost
three-twentieths of the entire length of the animal; the
caudal, about one-fourth of the same, has two large
sloping lobes, from the middle of the division of which
proceeds a longitudinal projection, extending over the
back as far as the dorsal fin. This tail, with the assistance of these vast lobes, forms a powerful lever, and

contributes mainly to the various evolutions of the animal.

The epidermis or skin is very soft to the touch, and of a deep blue or brilliant black superiorly, whitish underneath ; under the true skin is a thick bed of very white fat, convertible into oil.

The Dutch, Danes, and most of the marine people of the North, pursue this animal into remote and inhospitable regions, to obtain its fat. The Laplanders and natives of Greenland, whose taste is not the most remarkably delicate, feed on all parts of it, the flesh of which they boil or roast, having left it first to putrefy in order that it may become more tender.

The porpoise is frequently confounded with the dolphin ; however, it differs in having a shorter snout, which although somewhat sharply terminated is much narrower. In general, the porpoise is the smallest animal of the two, and rarely exceeds six or seven feet. It is of a thick form on the fore parts, tapering towards the tail like the other cetàcea.

The mouth is of a moderate width ; the teeth sharp, small, and numerous ; being commonly from forty to fifty in the jaw.

The porpoise lives chiefly upon the smaller fishes, and is observed, when in quest of food, to turn up the sand and mud at the bottom of the water like the common dolphin, and this uneasiness has been considered by seamen as a presage of stormy weather.*

---

* In the Morning Herald Newspaper of July 31, 1833, is the following article copied from the Dundee Constitutional.—" SALMON HUNT.— On Tuesday se'nnight, a great number of porpoises appeared off the Magdalene Yard in pursuit of salmon. It was astonishing to see the rapidity of their movements, their twistings and turnings in pursuit of their prey. The salmon were frequently observed to spring several yards out of the

According to the testimony of Fabricius, in his Fauna Gröenlandica, the D. Phoceana or Porpoise constantly swims in a curved posture, depressing very considerably both head and tail during that action. The period of gestation both in the porpoise and dolphin, according to Anderson, is six months, but Aristotle and Cuvier say it continues for ten months, and the female rarely brings forth more than one at a birth. She suckles her young with the utmost care, carrying it under the pectoral fins, which answer the purpose of arms : while the little one is yet feeble, she exercises it in swimming, sports with it, defends it with the utmost intrepidity until it can do without her cares. Their young grow rapidly; in ten years they attain their utmost length.

Porpoises are observed to assemble occasionally in vast numbers and to pursue shoals of herrings, mackerel, and other fish, which they drive into the bays with considerable velocity.

This animal was once considered as a sumptuous article of food, and is said to have been occasionally introduced at the tables of old English nobility; and this so lately as the time of Queen Elizabeth. It was eaten with a sauce composed of crumbs of fine bread with sugar and vinegar. However, it is now generally neglected even by sailors.

Notwithstanding the quickness with which they disappear under the water, numbers of them fall victims to the murderous skill of the fishermen. They have another enemy to fear not less redoubtable, one of their

water, but from the quickness of their enemies it appeared almost impossible that they could escape." The editor of the Dundee Paper considers the porpoises highly destructive to the Salmon Fishery.

own order, the *Physeter Microps*, or *Small-eyed Cachalot*, which pursues them with indefatigable perseverance, devouring them with amazing rapacity. Bingley states that one of these animals will yield about a hogshead of oil.

In America, the skin of this animal is tanned and dressed with considerable care. At first it is nearly an inch thick; but it is shaved down till it is quite thin and becomes somewhat transparent. It is made by the inhabitants into waistcoats and breeches, and said also to make an excellent covering for carriages.

The porpoise being the most common, and easier obtained than the other species of the cetàcea, has been more accurately examined, as to its structure and general characters.

The annexed engraving will give the reader some idea of the skeleton of this animal, and show its affinity to that of the Balæ'noptera Rorqual.

SPECIES II.

DELPHINUS DIDELPHIS,

OR

THE COMMON DOLPHIN.

THE body of the *Delphinus Didelphis* or Dolphin* is nearly oval. The figure of this animal has been greatly

---

* SYNONYMES.—*Le Dauphin ou Oye de Mer*. Belon. Hist. des Poissons. *Delphinus Antiguorum*. Willoughby and Ray. *Delphinus Corpore longo subtenti, rostro longo-acuto*. Artedi Synop. *Delphin*. Anderson. *Dolphin*. Borlase, Pennant and Shaw. *Dauphin Vulgaire*. La Cépède.

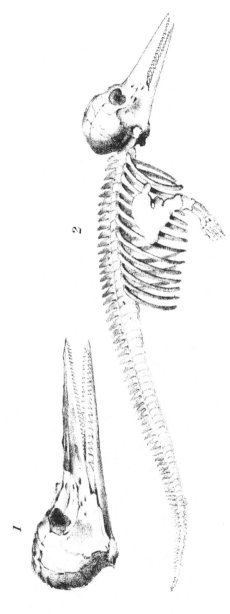

1. The Cranium of the Porpoise. | 2 The Skeleton of the Porpoise.

Fridal Lith. 24 Greek St.

misrepresented both by painters and sculptors. The recurved tail, the monstrous head with pendant lips, and eyes protected by enormous brows, which we see in the usual painted and engraved figures of a dolphin, are mere creatures of imagination.

The length of this species is frequently nine or ten feet, and its body about two feet at its thickest part, which is at the insertion of the pectoral fins, from which the body gradually diminishes towards the head and tail, and thus assumes an oval form, slender however in proportion to its length. The head enlarges at the top like that of the porpoise; but in this animal it diminishes in thickness, and ends in a muzzle, flattened something like the beak of a bird, whence the origin of Belon's name of *Oye de Mer*, or *sea goose*; but there is a transverse fold of the skin across the upper part of the snout. The jaws are of equal length, and furnished with a row of cylindrical teeth, a little pointed at one end, and projecting near one inch and a half above the gum. It would appear that the number of teeth varies according to the age and sex. Klein has counted ninety-six in the upper jaw and ninety in the lower. Mr. Pennant has, on the contrary, mentioned that he saw nineteen teeth in the latter, and twenty-one teeth in the former, by which I presume he means on one side. Other zoologists have observed forty-seven teeth in each jaw. In a specimen which in 1826 was exhibited in the streets of London, by a penny showman, I counted forty-six. This specimen was captured on the coast of Sussex, and measured seven feet nine inches in length.

The mouth of the dolphin is very wide, reaching almost to the insertion of the head. The dorsal fin is high, and placed rather nearer the tail, being curved backwards at its extremity, than the head; the swimming-paws or fins are situated low, being inserted at the under part of the breast, and are oval in shape. The tail is semilunar, and is composed of two lobes, the one of which folds over the other.

The upper surface of the body is black; the chest white. From beneath the eyes on each side passes a white ray, which stretches towards the pectoral fins. The dolphin is almost always an inhabitant of the open seas, and very rarely approaches the shore. It is commonly found in the Atlantic and Pacific Oceans, and is frequently seen accompanying ships on their passage to the East or West Indies. It occasionally appears in the British seas. The motions of this animal are inconceivably swift; hence it has been named by mariners as the *arrow of the sea.*

Like most of their congeners, they live upon small fish, though they will eat any offal and garbage that is thrown into the sea. They are said to attack whales; but I am of opinion that those who have stated this circumstance from having witnessed it must have mistaken another species for the dolphin. They are gregarious, and, like the porpoises, frequently sport about upon the surface, leaping out of the water, so as to be entirely visible. In these leaps, their back is a little curved, but not near so much as is commonly represented. They are said to change their colour before death; but this is probably an error, arising from a different reflection of the rays of light, when the body is in motion or at rest. On the pieces of money in circulation during the reign of Alexander the Great, and which are preserved by Belon, as well as on other medals, the dolphin is represented with a very large head, a spacious open mouth, and the tail raised above the head.

No animal has been more celebrated by the ancient poets

and historians than the dolphin. From the earliest ages he was considered as consecrated to the gods, and honoured as the benefactor of man. Pliny, Ælian, and other ancient authors, speak highly of his attachment to mankind. The younger Pliny has written a charming story of the love of a dolphin for Hippus;* and Ovid relates, with all the beauties of poetry, the story of the musician Arion, who, being pursued and thrown into the sea, was rescued and saved by this kind animal. He thus observes :—

> " But (past belief) a dolphin's arched back
> Preserved Arion from his destined wreck!
> Secure he sits, and with harmonious strains
> Requites his bearer for his friendly pains.
> The gods approve; the dolphin heaven adorns,
> And with nine stars a constellation forms."
>
> Ovid. Fasti. lib. ii. 117.

But after all these fabulous accounts of the dolphin by the ancients, and the presages drawn by modern seamen from their movements, it does not appear that this animal is endowed with more sagacity than the other species of the Cetàcea, or that it discovers greater attachment to man. To relate what has been said on this subject would far exceed my limits, and to no useful purpose, inasmuch as such of my readers as may feel inclined for a fuller examination will find it by referring to Pliny's Natural History, or to the writings of Athenæus and Ælian.

At present the dolphin is scarcely sought after as an object of traffic. Some centuries ago, however, its flesh was reckoned such a delicacy that we are informed, by Dr. Caius,† that one which was captured in his time was deemed worthy of being presented to the Duke of

---

* Plin. Epist. lib. ix. ep. 33.   † The founder of Caius College, Cambridge.

N

Norfolk, who distributed part of it among his most intimate friends. It was roasted, and served up with porpoise sauce. It is now very seldom eaten, except when
young and tender. The best parts are those next to the
head, the rest being dry and insipid.

SPECIES III.

DELPHINUS ORCA COMMUNIS,

OR

THE COMMON GRAMPUS. *

THIS is by far the largest animal of this genus. It is
the common grampus, and two varieties are to be met
with in the northern seas. They differ but little from
the delphinus phoceana, or common porpoise; in the
latter the snout is not so blunt, whilst that of the
grampus is short, and a little turned up. The remarkable difference of size too is very striking, the grampus
being from twenty to twenty-four feet long, and proportionally bulky. The latter also is furnished only with
forty teeth, whilst the former has forty-six in each jaw.

The grampus is of an extremely fierce and pre-
daceous disposition, feeding on the larger fishes, and
even on the dolphin and porpoise. It is also said to
attack whales and to devour seals, which it occasionally
finds sleeping on the rocks, dislodging them by means
of its back fin, and precipitating them into the water.
This animal is found both in the Mediterranean and
Atlantic seas, as well as in the polar regions. The

---

* SYNONYMES. — *Delphinus Orca Communis,* DEWHURST; French,
Epaulard—Spek-Hugger; Norwegian, *Hoval-Hund;* Dutch, *Botskop;*
Icelandic, *Huyding;* Swedish, *L'Opare. L'ondrean Grand Marsoin,*
Belon. *Balæ'na minor utraque maxilla dentata,* Sibbald. *Grampus,*
Pennant and Shaw. *Dauphin orque,* La Cépède.

Delphinus Orca Communis
or Common Grampus

lower jaw is much wider than the upper, and the body somewhat broader and deeper in proportion: the dorsal fin sometimes measures not less than six feet in length from the base to the tip. Fabricius emphatically styles it *Balæ'narum Tyrannus*, and it is considered as one of the most ferocious inhabitants of the ocean.

The following is a description of a specimen of this species of grampus, which was captured in Lynn harbour, about ten miles from the town. The animal, which was a male, was discovered by some fishermen on the 19th of November, 1830, with his dorsal fin rising just above the surface of the water. He was immediately driven into the shallows, and attacked by the men; but, not being provided with proper weapons, it was with much difficulty they were able to despatch him by the help of knives and sharpened oars. The groans of the poor animal are described as having been most horrible, and the effusion of blood very great. Being at length deprived of life, he was towed up the river to the town, and landed on the quay, whence he was drawn by six horses through the streets, to the place where the carcass was flensed, or cut up.

The accompanying drawing * is a faithful representa-

---

* Obligingly lent me for this work, by J. C. Loudon, Esq., F. R. S., &c., Editor of the Magazine of Natural History, from which this account is extracted. Vide vol. iv. p. 341.

tion of the body when lying on the ground; and the
dimensions, taken as accurately as circumstances would
admit, were as follows:—

	Feet.	Inches.
From the tip of the upper jaw to the division of the tail, following the curve of the back	21	3
From the tip of the under jaw to the same in a straight line	18	11
From the tip of the under jaw to the anterior edge of the dorsal fin	8	2
From the posterior edge of the dorsal fin to the division of the tail	10	9
Base of the triangular dorsal fin	2	4
Height of ditto.....................................	4	0
Length of the ovate pectoral fin....................	3	11½
Width of ditto......................................	2	8
Distance of the two lobes of the caudal fin............	7	1
Circumference of the body	14	0
From the tip of the upper jaw to the spiracle	2	7
From the tip of the lower jaw to the anterior edge of the pectoral fin...	4	0
From the back tooth of the lower jaw to the same on the opposite	0	10

The teeth are twenty-four in each jaw; the seven
backward ones are cuspidate: the rest appear to have
been the same, but are now worn down. Those in the
front are nearly concealed, and a few of the central ones
are curvated. The upper jaw projects a little beyond
the lower. The eye is five inches above the corner of
the mouth. The orifice of the ear is scarcely large
enough to admit a pea, and is placed just behind the
eye in the white spot. The spiracle was so much injured by the fishermen, in their attempts to kill the
animal, that its form could not be correctly ascertained;
but it appears to have been lunated, with its horns
turned towards the nose. The skin is a fine glossy
black; but the under jaw, the belly, a singular oval spot
behind each eye, and a large mark on each side, are of a

pure white; the saddle-like mark on each side the back is of a silvery grey. The dorsal line is prominent, sharp, and of a jet black. Its weight is computed at about three and a half tons; and it was purchased of the fishermen for £23; it was not, however, expected to yield a sufficiency of oil to repay the purchaser; but as he realized something by exhibiting the animal, and afterwards sold its head for £7 to a gentleman in the neighbourhood, he was not likely to be a loser by his bargain.*

SPECIES IV.

DELPHINUS GLADIATOR,

OR

THE SWORD GRAMPUS.†

THIS species has the dorsal fin long and bony, broad at the base, and curved like a scymitar. As these animals advance in age, this instrument grows longer; so that the leader, or old animal, can be distinguished from his followers by the superior height of his fin. This is one of the fiercest enemies of the whale, from his being thus provided with such an efficient weapon of annoyance.

The sword grampus pursues also seals, and the latter, in their clumsy efforts to escape upon the ice or rocks, are frequently overtaken by their active adversary. When the seals are swept from their place of retreat back into the water, they are speedily vanquished.

The sword grampus varies much in size according to

---

* The author of this paper signs his communication G. M., Lynn Regis,—December 1, 1830.

† SYNONYME.—*Dauphin Orque avec un sabre.*

its age; but when full grown, as the one exhibited in the engraving, it is then above twenty feet long. The great size of the dorsal fin, from which the animal derives its name, distinguishes it among the dolphins as much as a similar organ does the *Physeter Tursio* (or finned Cachalot) among the Cachalots.

These animals are found in Davis's Straits, on the coasts of America and Spitzbergen, in troops from six to eight. They generally live on fish, but the young Balæna Mysticete is assailed on all sides by them. This peaceable animal is tormented, harassed, and even forced to succumb to the attacks of these audacious adversaries; for, when he opens his mouth for the purpose of respiration, in a moment these grampuses seize upon his tongue with the greatest fury, and tear it to atoms.

The grampuses have frequently been captured in the River Thames.—In 1759 and 1772, two common grampuses were taken; and in 1793 six of the sword species were caught in the same river, one of which, after making considerable resistance, was ultimately destroyed near Greenwich: it measured thirty-one feet in length, and twelve feet in circumference.

SPECIES V.

DELPHINUS BIDENTATUS,

OR

THE TWO-TOOTHED DOLPHIN.*

THE body of this species is conical. The dorsal fin is spear-shaped. There are two sharp teeth in the lower

---

* SYNONYMES.—*Le Dauphin a Deux Dents*, Bonnaterre. *Delphinus Bidentatus*, La Cépède.

jaw. In many of its characters it resembles the *Delphinus Tursio,* but in others it is so different that it may very properly be considered as a distinct species. The forehead is convex and rounded. The snout is slender and flat; and the upper jaw is flat and ends in a beak somewhat like that of a duck. The teeth are placed at the anterior extremity of the lower jaw. The pectoral fins, which are of an oval form and small for the size of the body, are placed opposite to the angles of the mouth. The place of the dorsal fin corresponds to the origin of the tail, is spear-shaped, pointed, and inclines backwards. The tail-fin is divided into lobes, forming by their union a crescent. The lower part of the body is of a light brown colour; the upper part is brownish-black. This species is supposed to be from thirty to forty feet long.

SPECIES VI.

DELPHINUS FERES.

In this species there is one fin on the back. The head is rounded and nearly of the same height as the length. It is very thick at the top, and, suddenly diminishing towards the anterior part, ends in a short round snout. The jaws are equal in length; they are covered with membranous lips, and are furnished internally with a row of oval and obtuse teeth: twenty have been reckoned in each jaw. The form of the teeth constitutes the distinctive character of the species. The large and small teeth are equal in number. The largest are above an inch long, by half an inch broad. The small teeth are only five or six lines in length.

A skeleton of one of this species is preserved in the Cabinet of Natural History at Frejus, in France. The length is fourteen feet. The bones of the skull are one foot ten inches long and one foot five inches broad.

It is found in the Mediterranean Sea, and occasionally in the Arctic Ocean.

SPECIES VII.

DELPHINUS TURSIO.*

THE form of the body is conical; the dorsal fin is curved; the snout is compressed above; the teeth are straight and blunt. The greatest thickness of this species is between the dorsal and pectoral fins. From this to the extremity of the tail the body becomes more gradually slender. The breathing-hole is placed above the orbits of the eyes; it is about one inch and a half in diameter. The anterior part of the head is inclined and rounded, and terminates in a flat beak. The lower jaw is the longest. Both jaws are furnished with forty-two cylindrical teeth, which are disposed in a single row.

The pectoral fins are situated very low. The dorsal fin rises like an inclined plane and is incurvated behind. At the posterior base of the latter fin there arises a projection stretching to the tail. The tail-fin is divided into two lobes in the form of a crescent. The upper part of the body is black, and the belly white.

It has been noticed by many zoologists that when this species rises to the surface to respire a great part of the

---

* SYNONYMES.—*Delphinus Nésarnak*, Bonnatérre. In Greenland it is denominated the *Nésarnak*.

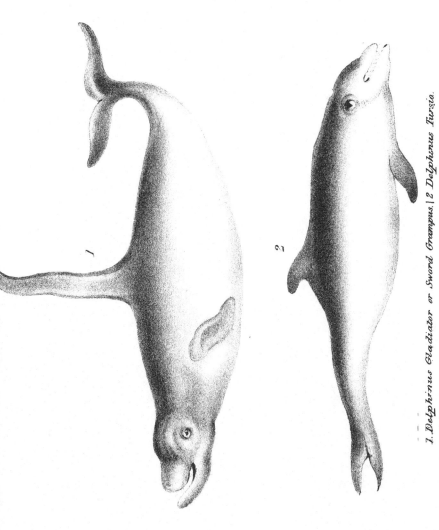

1. *Delphinus Gladiator or Sword Grampus.* | 2 *Delphinus Tursio.*

body appears above water.   It inhabits the open seas, and is consequently captured with extreme difficulty. Its flesh has been eaten in the same way as the porpoise, but now is disused.

GENUS VII.

HYPERÖODON.

SPECIES.

HYPERÖODON BUTSKOPF, vel DELPHINUS DEDUCTOR,

THE BOTTLE-NOSED, OR CA'ING WHALE.*

Dr. TRAILL, one of the Professors in the University of Edinburgh, has ably described this species of the *Genus Delphinus,* and which was considered as a new species of whale by this gentleman. Ninety-two of them were stranded in Scalpay Bay, in Pomona, one of the Orkney Islands, a few days prior to the great storm in December, 1806.   On the 21st of February, 1805, one hundred and ninety of these animals, varying in length from six to twenty-four feet long, were forced ashore in Uyea Sound Unst, West Shetland; on the 19th of

---

SYNONYMES.—*Delphinus Orca, (Butskopf),* Linn. *Butskopf,* Martin. *Bottle-head* or *Flounder's-head,* Dale. *Nebbe Hual,* Pontopp. Norway. *Beaked Whale,* Pennant. *Dauphin Butskopf,* Bonnatérre. *Hyperöodon Butskopf,* La Cépède. The *Ca'ing Whale* of the Shetland and Orkney Islands.

Delphinus Deductor; or Bottlenosed Whale of Shetland.

Delphinus Bidens; or Two Toothed Dolphin.

Friedel. Lith. 24. Greek Street, Soho.

March following, one hundred and twenty more came to the same spot, in all three hundred and ten.   In this second shoal there were probably about five hundred, but very many escaped.*

This animal, observes Dr. Traill,† clearly belongs to the genus *Delphinus*, of the class *Mammalia*.   The only hitherto described species of that genus which it at all resembles is the *Delphinus Orca*, or Grampus; but it is distinguished from the latter by the shape of its snout, the shortness of its dorsal fin, the length and narrowness of its pectoral fins, the form and number of its teeth, and the colour of its belly and breast.

It abounds in the seas around the Orkney, Shetland, and Feroe Islands, and even in Iceland.   Being of a gregarious disposition, the main body of the drove follow the leading whales, as a flock of sheep follow the wedders.   This disposition being well known to the natives of Shetland and the Orkneys, and is improved to their advantage; for, whenever they are enabled to guide the leaders into a bay, they are sure of likewise entangling great numbers of their followers.‡   From this peculiarity of following a leader, Dr. Traill suggests the name of *Deductor*, instead of the name he formerly gave.   The Shetlanders gave them the name of Ca'ing or Leading Whales.

The whole body is almost black, smooth, and shining like oiled silk.   The back and sides are jet black, the belly and breast of a somewhat lighter colour.   The general length of the full grown ones is about twenty feet.   The body is thick.   The dorsal fin does not ex-

---

* Tour through some of the Islands of Orkney and Shetland, by P. Neill.   Edin. 1806. p. 221.
† Nicholson's Journal.   Vol. xxii. p. 81.
‡ Neill's Tour, p. 222.

ceed two feet in length, and is rounded at the extremity. The pectoral fins are from six to eight feet in length, narrow and tapering at their extremities. The head is obtuse; the upper jaw projects several inches over the lower in a blunt process. It has a single spiracle. The full grown have twenty-two subconoid teeth, a little hooked. Among those stranded in Scalpay Bay were many young ones, which, as well as some of the oldest, wanted teeth. The youngest measured about five feet in length, and were still sucklings. The females had two teats, larger than those of a cow, out of which milk flowed when they were squeezed.

These animals frequently enter the bays around the Shetland and Orkney coasts, in quest of small fish, which seem to be their food. When one takes the ground, the rest surround and endeavour to assist it: from this circumstance several of them are generally taken at once. Dr. Traill has frequently observed an animal, which he conjectures to be of this species, elevating its dorsal fin and a considerable part of its back above the waves, with a slow tumbling motion, for many successive times. They are inoffensive and rather timid. They are chased on shore not unfrequently by a few yawls, and seen to follow one as a leader with blind confidence. Dr. Traill was once in a boat when the attempt was made to drive a shoal of them ashore; but, when they had approached very near the land, the foremost turned round with a sudden leap, and the whole rushed past the crew, with considerable velocity, but carefully avoided the boat.

These animals are extremely fat, and yield a considerable quantity of good oil.*

---

* Nicholson's Journal, vol. xxx. p. 82.

In April and August 1824, during the time we lay at anchor in Brassa Sound (Lerwick), I observed great numbers of them, as well as during the voyage to and from the icy seas. Captain Scoresby in his work has copied, from a volume lent him by Dr. Traill, a description of the mode of capturing the *Deductor* by the natives of the Feroe Islands in the 17th century. From this work it appears that numerous animals of this kind, called in Feroe the *guard-whales*, were frequently driven ashore by the boats, and killed, and that in the year 1664 there were taken at two places about a thousand. Many other historical notices of the capture of these shoals are to be met with. In the year 1748, forty of them were seen in Torbay, and one seventeen feet in length was killed. In 1799 about two hundred of them, varying from eight to twenty feet in length, ran themselves ashore at Taesta Sound, Fellar, one of the Shetland Isles. In the winter of 1809 and 1810, one thousand one hundred of these whales approached the shore in Hvalfiord, Iceland, and were captured; and, in the winter of 1814, one hundred and fifty of the same were driven into Balta Sound, Shetland, and there killed. There are only instances of a very small proportion, of which in modern times, an extensive slaughter of the Delphinus Deductor has taken place on the shores of the British and other northern islands.

In addition to these, the following extract from the Caledonian Mercury may not be uninteresting:—

" The thriving little town of Stornaway, in the island of Lewis, was on Wednesday week (April 25th, 1832) enlivened by a scene of the most animated and striking description. An immense shoal of whales was, early in the morning, chased to the mouth of the harbour by two fishing boats, which had met them in the offing. This

circumstance was immediately descried from the shore,
and a host of boats, about thirty or forty in number,
armed with every species of weapon, set off to join the
others in the pursuit, and engage in the combat with
these giants of the deep. The chase soon become one
of bustle and anxiety on the part both of man and whale.
The boats were arranged by their crews in the form of
a crescent, in the fold of which the whales were collected,
and where they had to encounter tremendous showers
of stones, splashings of oars, frequent gashes with harpoon and spear, whilst the din created by the shouts of
the boats' crews and the multitude on shore was in itself sufficient to stupify and stun the bottle-nosed foe to
a surrender. On more than one occasion, however, the
floating phalanx was broken, and it required the greatest
activity and tact ere the breach could be repaired, and
the fugitives regained. The shore was neared by degrees, the boats advancing and retreating by turns, till
at length they succeeded in driving the captive monsters
on a beach opposite to the town, and within a few yards
of it. The gambols of the whales were now highly di-
verting, and, except where one became unmanageable
and enraged when a harpoon was fixed, or his tail entangled in the noose of a rope, these creatures were
not dangerous to approach. In a few hours the whole
were captured. The shore was strewed with their dead
carcasses, whilst the sea presented a bloody and troubled
aspect, giving evident proofs that it was with no small
effort that they were subdued and made the property of
man. For fear of contagion, the whole, amounting to
ninety-eight in number (and some of them very large),
were towed immediately to a spot distant from the town,
and sold by public roup (auction), the proceeds arising
from which to be divided amongst the captors. An

annual visit is generally paid by this species of the whale tribe to the island of Lewis, and, besides being profitable when captured, generally furnish a source of considerable amusement. On the present occasion, the whole inhabitants of the place, both male and female, assembled to the scene of slaughter, where they were evidently delighted spectators at the scene of death! and occasionally rendered assistance. It is to be lamented, however, that a young sailor received a stroke from the tail of the largest of the whales, and but little hopes are entertained of his recovery."*

GEÑUS VIII.

DELPHINAPTERUS.

SPECIES.—DELPHINAPTERUS BELUGA,

OR

THE WHITE WHALE †

This is a species which appears not to have been very distinctly known until within the last half century. It is a native of the Northern Seas, the Arctic Icy Sea, and especially in Hudson's Bay, and Davis's Straits. When

---

* It is to be regretted that the Editor of the Caledonian Mercury did not give his readers any idea of the dimensions of these creatures, or the amount they sold for, and the quantity of oil they produced.

Since the above, the newspapers have recorded several similar shoals of this species of the *Genus Delphinus.*

† SYNONYME—The *Beluga Whale* of Pennant and Shaw.

Delphinopterus Beluga.—or White Whale.

the winters are excessively rigorous, it is reported that it quits these icy climates in search of more temperate regions, and occasionally ascends rivers. These animals are very gentle, and so familiar that when they discover a vessel they swim away in a crowd to meet, forming a kind of suite, and playing all manner of gambols.* The Beluga, however, is no object of research to the whalers, who set very little store by it. Mariners assert that it is usually the precursor of the common whale, and that whenever it makes its appearance this monarch of the cetàcea is at no very great distance.

Sometimes they are captured either by very strong nets or harpoons, for the sake of the oil they produce; when nets are used, they are extended across the stream, so as to prevent their escape out of the river; and, when thus interrupted in their course to sea-ward, they are attacked with lances, and great numbers of them are sometimes killed. Captain Scoresby states that he has seen them several times on the coast of Spitzbergen, but never in numbers of more than three or four at a time.

The attachment of the female for the single young one is very great, and would even be an honour to human beings; she retains it continually by her side, and is never separated from it without displaying the most lively un-easiness. She frequently presents it to one of the mammæ or breasts, which are situated near the tail, and the little one sucks with avidity; it is of a dazzling white-ness, and, to use the language of Dr. Shaw, the whole forms a beautiful spectacle.

The Beluga subsists upon cod, pleuronectes* (or flat-

---

* Cuvier, Régne Animal, vol. iv. p. 458.

* This signifies a genus of flat-fish, of which the sole, flounder, plaise, turbot, halibut, &c., form species.

fish), and a variety of other fish, which it swallows with such avidity as often to risk being suffocated, the orifice of the throat is so narrow; the flesh is said to have a reddish tint.

Fabricius, Dr. Pallas, and Baron Cuvier, have accurately described this, which is the most elegant of the whole tribe, and when full grown is entirely milk-white: in some specimens they are tinged very slightly of a rose-colour, and in others of a bluish cast.

The head of the Beluga is small and elongated. The anterior part of the body presents the figure of a cone, the base of which may be considered to rest about the pectoral fins against that of another cone much larger, and composed of the body and tail; the muzzle is elongated and rounded in front; mouth not very wide; both jaws are equal, and furnished with nine or ten small teeth, blunt at the top, but unequal and distinct from each other; the largest are in front of the termination of the muzzle, where the mouth, small in proportion to the size of the animal, is situated.

Dr. Shaw states it to measure from twelve to eighteen feet in length, Baron Cuvier from eighteen to twenty-one feet, and Count La Cépède six or seven metres.

Instead of a dorsal fin, the Beluga has only a sort of longitudinal projection on the back, which is semi-callous, and seems to possess but little sensibility. The pectoral fins are oval, broad, and thick; the longest of the phalanges or fingers, which are all enveloped in a membrane, has five articulations. The skin, as already stated, is of a whitish colour when the animal is young, with a multitude of brown spots.*

---

* I am afraid that Baron Cuvier errs with respect to the brown spots, inasmuch as neither Fabricius nor Captain Scoresby has mentioned them,

Above the anterior part of the head is a protuberance which forms the common orifice of the spiracles; its situation in the head causes the water rejected by the Beluga to fall behind it.

The eye is small, round, projecting, and of a bluish colour; at some distance behind it, is the meatus auditorius or opening of the external ear, but the orifice is so small as almost to be imperceptible.

The skin is on every part very smooth and slippery, and in general the Beluga is very fat.

The Beluga is not unlike the narwhale in its general form, but is thicker about the middle of its body in proportion to its length.

A male animal of this species was captured in the Firth of Forth, in the month of June 1815. Its length was thirteen feet four inches, and its greatest circumference nearly nine feet.

When this animal swims, says Dr. Pallas, it bends the tail inwards in the same manner as a crawfish, by which means it possesses the power of swimming extremely fast, by the alternate incurvation and extension of that part. Dr. Shaw says it has so great an affinity for the *phocæ* or *seals*, that the Samöides consider it a kind of aquatic quadruped. It produces only one young at a birth, which is at first of a blue tinge and sometimes grey, or even blackish; acquiring as it advances in age a pure milk-white colour.

---

and were it so, I think that the last gentleman who has been familiar with them would have mentioned this circumstance. He has seen some of a yellow colour, approaching to an orange.

CONCLUDING OBSERVATIONS ON
THE CETACEA.

———

Having finished my remarks on each individual species of the cetàcea, I may not improperly conclude with a few general recapitulatory observations on this tribe in general.

ABODE OF THE WHALE, &c.

According to the testimony of the ancient naturalists, the whale was more frequently seen in the ocean than at present; for, on account of its being disturbed by the numerous fleets traversing the ocean, they have retired to the northern regions, where they are less exposed to the noise of the mariners, less harassed by the fishermen, and enjoy that tranquillity which is no longer found in their former haunts.

The B. Mysticètus is most frequently found in the Greenland seas, Davis's Straits, the coast of Spitzbergen, Iceland, and Norway; on the coast of Labrador, in the Gulf of St. Lawrence, and round Newfoundland, by the bays of Baffin and Hudson, in the sea to the northward of Behring's Straits. It is also found among the Philippine islands, near Socotora (an island on the coast of Arabia Felix), and on the coast of Ceylon. It likewise frequents the Chinese seas; and, if the reports of voyagers are to be believed, it is found there of immense size. The usual retreat of the sperm-aceti whale is the northern ocean, towards Davis's Straits, North Cape, and the coasts of Finmark. Of all the cetàcea this appears to lead the most wandering life. In 1787, great numbers of this species were discovered in an extensive bay on the southern peninsula of Africa, at the distance of forty leagues from the Cape of Good Hope.

It is hardly ever met with in the German Ocean, and rarely within two hundred leagues of the British coasts; but along the coasts of Africa and South America it is met with periodically, in considerable numbers. In these regions it is attacked and captured by the southern British and American whalers, as well as by some of the people inhabiting the coasts in the neighbourhood. Whether this whale is precisely the same kind as that of Spitzbergen and Greenland has not yet been ascertained, although it evidently is a *mysticètus;* perhaps it may be some important variety of this species. However, there is one striking difference, which may possibly be the effect of situation and climate, which is, that the balæna mysticètus which is found in the southern regions is frequently found covered with barnacles, vulgarly called the *whale-louse,* whilst those of the Arctic Seas are free from these shell fish.

The fishing grounds for the spermaceti whale are—1st. From the Sechelles Isles (belonging to us), to Timor, and all the coast of New Holland, as far as Shark's Bay. 2nd. The Japanese seas, as far as the Philippine Isles, and to the eastward as far as California. The black whale comes from the south polar seas in May to bring forth its offspring; it remains in the bays of New Holland, Africa, and South America, till August, and on the coasts till November, when it returns in a south-westerly direction. The possession of King George's Sound, to the southward of the Swan River settlement, affords us immense advantages and facilities for carrying on these lucrative branches of industry, as it enables our vessels to fish for the spermaceti as well as the black whale in the same year.*

The dolphin genus is found in all seas, in the Arctic Ocean, the Mediterranean, the Gulf of Messina, and the Adriatic Sea, from whence they go into the lagoons of Venice, and to the coasts of Cochin and China. Very considerable establishments are formed for their capture, which produce a great quantity of oil.

I may conclude by stating that the *Balæna Mysticè-tus,* or Greenland whale, and the *Monodon Monoceros,* or Narwhale, usually frequent the seas towards the poles, between the 68th and 80th degrees of latitude; and that the other families are found diffused more or less in the seas of more temperate regions. It would appear from the preceding description of the places forming the ordinary haunt of the whale, that the productions of nature are disposed somewhat in a contrary order, since we find all the large terrestrial animals, such as the elephant and rhinoceros, in countries within the torrid zone; while the huge inhabitants of the ocean have fixed their abode in the polar regions.

MIGRATION OF THE WHALE.

Although the abode of the whale be generally determined and fixed, yet particular causes force them to

---

* In the Metropolitan Magazine for April, 1833, the reader will do well to peruse an excellent article on "*The South Seas,*" inasmuch as it contains some valuable commercial observations, respecting the capture of common and sperm whales.

leave their usual and natural haunts. The *season* of their amours, a furious storm, the pursuit of a harassing enemy, the want of food, or excessive cold, often obliges them to migrate. Sometimes they appear solitary, sometimes in considerable numbers, according to the nature of the causes which have disturbed and driven them from their ordinary retreats. According to the information of voyagers who have visited these regions, the great whale every year, in the month of November, leaves Davis's Straits, enters the river St.Lawrence, and there brings forth her young, between Camarca and Quebec; and from thence, in the month of March following, they regularly return to the polar seas.

It appears that the whale constantly remains in the northern ocean, never leaving it till the female brings forth her young, or when driven away by an enemy. In this last case they are found generally solitary, at least not more than the male and female, or the mother and the young one. The spermaceti whales, however, seem frequently to change their habitation, and to roam about in strange seas, from considerable numbers having been thrown ashore, or left dry by the retreating tide, at different times. In the year 1690, two hundred of this species were landed near Cairston, in the Orkneys; and in 1784, thirty-one large spermaceti whales came on shore on the west coast of Audirne, in the province of Lower Brittany, in France.

ENEMIES OF THE WHALE.

The greatest and most terrible of the small whale is the physeter microps, or small-eyed cachalot. As soon as he perceives the pike-headed whale, the porpoise, and some others, he darts upon them, and tears them to pieces with his crooked fangs. It is said there exists a continual and settled enmity between the narwhale and the great whale; and that they never meet without engaging in combat, in which the whale receives so many severe, and often deadly wounds, frequently occasioning death. When the narwhale strikes its tooth or horn into the side of ships, it is supposed that it is through its mistaking the vessel for its enemy the whale. The white bear (*Ursus Borealis* vel *Polaris*), so common in Greenland and Spitzbergen, is particularly fond of the flesh of cetàceous and other oceanic large animals. He remains constantly on the watch for his prey on a mass of ice, or on the sea shore, and as soon as he perceives it he throws himself into the water, and plunges to attack it. The large and the small whales are equally the objects of his eager pursuit; but he is not successful till after they have lost a great deal of blood from the wounds which he has inflicted, or they have been exhausted with fatigue.

Between the sword-fish and the whale there constantly exists a warfare. It is related by all the whalers that the whale and sword-fish, whenever they meet, join in combat, the latter being always the aggressor. Sometimes two or more individuals combine to attack a single whale; and it is inconceivable with what fury they make the attack. The whale, whose only defence is his tail, endeavours to strike his enemy with it, and a single blow would prove mortal. But the sword-fish, with astonishing agility, shuns the dreadful stroke, bounds into the air, and returns upon his huge adversary, plunging the rugged weapon with which he is furnished into his back. The whale is still more irritated by this wound, which only becomes fatal when it penetrates through the fat. The engagement ceases not but with the death of one of the combatants. Martens relates an account of one of these combats between the Iceland whale (*Balæna Icelandica*) and the sword-fish. It seemed to be extremely dangerous to approach the field of battle. It was, therefore, at some distance that he saw them pursuing and striking each other, dealing such violent blows, that the water rose in foam as if agitated by a storm. He was prevented seeing the issue of the fight by the weather becoming thick and hazy;* but he was informed by the sailors, that such combats were frequent; that they generally kept at a distance till the whale was vanquished; and that the sword-fish, only eating the tongue, relinquished the rest of the body, which they take possession of, and I need hardly say, are glad to obtain it.

Forskal informs us, that the Arabians believe that some species of the *scarus*, a fish found in the Red Sea, enter the spiracles of the whale, and destroy it with their sharp spines; and, in confirmation of this fact, it is mentioned, that one of these fishes was found in the spiracles of a dead whale.

The whale is even harassed with aquatic birds, which alight in great numbers on his back, in search of the testaceous animals and small insects, which have made it their habitation. And like most other animals the whale is tormented with a species of louse (*Oniscus Ceti*. LINN.), peculiar to itself, which adheres so strongly to the skin, that it will sooner be torn asunder than be compelled to let go its hold. The fins, the lips, the parts of generation, and other parts of the body which are most protected from friction, are chiefly infested with this parasitical insect. The bite is ex-

---

* Vide page 89.

tremely painful, and they are most troublesome in that season when the whale is in heat.

AGE OF THE WHALE.

If the time necessary for the growth or increase of the body were in proportion to the period of life, there could be but little doubt of the whale being, of all animals known, the most remarkable for longevity. It is well known, that the whales which were taken when this fishery first became an object of trade, that is, between two and three hundred years ago, were of much greater bulk than those found in the present day. The largest now taken rarely exceeds sixty feet long; while, at that time, some reached a hundred in length. The reason of this difference seems to be, that when the fishery first commenced, whales having met their perfect growth were to be met with. These, on account of being the largest, were constantly harassed, pursued, and destroyed; so that none which have obtained their full growth are now to be found in those seas. From this circumstance, that no large whales are now to be seen in the places which they commonly frequent, it is concluded, that the period of the life of the whale is very long; and that they cannot arrive at the huge size for which the first whales were so remarkable, since they are not permitted to live undisturbed the requisite length of time to attain that bulk. According to Buffon and Cuvier, a whale may live one thousand years, since a carp has been known to reach the age of two hundred. But, reasoning from analogy, with regard to the structure and economy of the whale, we have seen in many instances by no means holds; and it is perhaps equally inapplicable to the growth and age of this order of animals.

PRESENT STATE OF THE BRITISH WHALE-
FISHERY.

I shall conclude with a few notices of the present state
and prospects of the British Whale Fishery, considered
in a commercial point of view. For the particulars I am
about to mention, I am principally indebted to Mr.
Macculloch's Dictionary of commerce, and to the volume
I have already mentioned of the Edinburgh Cabinet
Library, containing, I believe, the latest account of the
fishery that has yet appeared.

According to Captain Scoresby, the average quantity
of shipping fitted out for this trade for the nine years
ending with 1818, in all the English ports, *viz.* London,
Hull, Whitby, Newcastle, Liverpool, Berwick, Grimsby,
and Lynn, was $91\frac{5}{9}$ vessels; and in the Scotch ports, *viz.*
Aberdeen, Leith, Dundee, Peterhead, Montrose, Banff,
Greenock, Kirkaldy, and Kirkwall, $40\frac{1}{9}$. In 1830 the
former quantity had diminished to 41; while the latter
had only increased to 50. Upon the whole, therefore,
there has been a falling off in the course of twelve years
to the extent of about 30 per cent. The season of 1830
was one of the most disastrous ever known since the
commencement of the fishery. Of the ninety-one
vessels which sailed, nineteen were entirely lost; as
many more returned clean (or without a single fish);
seventeen brought only one fish each; and of the others
many had only two or three. The actual loss incurred
from the shipwrecks, and the severe injuries sustained
by twelve other vessels, was calculated to have amounted
to about 143,000*l.* Both oil and whalebone immediately
rose to more than double their former price; but still
the whole produce of the fishery of this year did not

amount, according to the highest estimate, to more than 155,565*l.*; while that of 1829 was reckoned at 376,150*l.* The season of 1831 was also unfortunate, though not to the same extent; three of the vessels having suffered shipwreck. The produce as compared with that of the preceding year was, in oil, 4800 tons in place of 2205, and of bone (baleen) 230 tons in place of 119. But in 1829 there had been obtained 10,672 tons of oil, and 607 tons of whalebone; and in 1828, of oil 13,966 tons, and of bone 802 tons. The value of the whole produce of the fishery of 1831, when oil had fallen from 50*l.* to 30*l.*, and whalebone from 380*l.* to 200*l.*, was estimated only at 190,000*l.* The season of 1832 was considered prosperous.

It would be unfair, however, to judge of the value of the trade entirely from these two years. " The British fishery," it is remarked by the writer in the Edinburgh Cabinet Library, " has lately yielded a produce and value much exceeding that of the Dutch, even during the period of its greatest prosperity. In the five years ending with 1818, there were imported into England and Scotland 68,940 tons of oil, and 3,420 tons of whalebone; which, valuing the oil at 36*l.* 10*s.* and the bone at 90*l.*, with 10,000*l.* in skins, raised the entire product to 2,834,110*l.* sterling, or 566,822*l.* per annum. The fishery of 1814, a year peculiarly fortunate, produced 1437 whales from Greenland, yielding 12,132 tons of oil, which, even at the low rate of 32*l.*, including the whalebone and bounty, and added to the produce from Davis's Straits, formed altogether a value of above 700,000*l.*"*

---

* The price of baleen or whalebone at the present period(August, 1833), I am informed is about *two shillings and two-pence* per pound, *retail*, but that the *wholesale* or market price is 160*l.* per ton.

The whale trade has also been gradually shifting from the ports in this country, which formerly enjoyed the greatest share of the business. Previous to the year 1790, London was the principal port from whence the vessels sailed, inasmuch as four times the number were sent out from the Greenland Docks,* at Deptford, than from any other place. Even in 1820 the metropolis of England continued to send out about seventeen or eighteen which were engaged in this trade; in 1824, when I sailed, there were but four, the Neptune (my own ship), belonging to Messrs. Benson and Hunter, of Shadwell, the Dundee, the property of Messrs. Gale, with the Rookwood, and another vessel (whose name I have forgot, but I believe it was the Margaret), the property of William Mellish, Esq.† At the present period I believe there are but two belonging to the same gentleman. Liverpool, in a similar manner, after having for a considerable period carried on the trade to a very considerable extent, has now entirely relinquished it. Whitby, also, which sixty or seventy years ago was largely engaged in it, now sends out only one, or at the most two ships. We may now very properly consider Hull as the principal whale-fishing port in Great Britain, and it has been so since the commencement of the present century. In 1830 that town sent out thirty-three ships. Peterhead on the eastern coast of Scotland may be ranked next to Hull, having sent out thirteen ships in the same year. After it may be ranked Aberdeen, Dundee, Leith, Kirkaldy, and Burntisland. In Peterhead, and in fact most of the Scotch ports, the trade

---

* They are now denominated the Commercial and East Country Docks, and are situated at the end of the Lower Road, about three-quarters of a mile from the town of Deptford.

† The same gentleman whose life was recently attempted by a maniac.

may be considered to be on the increase; this I account for, from the circumstance of what few whales are captured being solely confined to these ports, instead of being diffused among those already mentioned.

The most profitable period for the whale-fishery was during the period of the bounty, which it is calculated cost the nation from 1750 to 1824, upwards of two millions and a half, and it is to the eternal disgrace of those in power, who took it off, by so doing thereby preventing the exercise of that spirit of enterprise and adventure so characteristic of British merchants and seamen. And many of the sums now improperly spent and conferred upon idlers, might with greater benefit to the nation be thus appropriated, and the time is not far distant, when too late, this will be discovered, if to many reflecting minds it is not already. The bounty stimulated the merchant to equip his vessel: if she was unsuccessful in the voyage, his loss was lessened, whereas now the whole burden falls upon his shoulders; consequently, the oil rose in price, and scientific men having produced coal-gas at a cheaper rate, and at the same time producing a more brilliant light than could be afforded by oil, it has become among tradesmen an almost universal substitute. These combined circumstances have produced a less demand for oil, merchants employ their vessels for other purposes, and the consequences will be, that the whales will have a better chance to multiply, and the trade will be thrown into the hands of the enterprising Americans, who richly deserve it.

PART II.

THE

NATURAL HISTORY

OF

THE OCEANIC INHABITANTS

OF THE

ARCTIC REGIONS.

I now proceed to describe the remaining inhabitants of the
Arctic Seas, as far as they have been discovered; and shall
therefore commence with pointing out the zoological peculiarities of those fishes which frequent, generate, and emigrate
from the Icy Ocean. Ichthyologists as yet are unacquainted
with the number of species residing in this quarter of the
mighty deep; but, as far as I have been able to ascertain, they
are as enumerated in the following descriptions, but no doubt
succeeding zoologists will discover many more, whose existence in the frozen seas are not even conjectured.

THE SEA SERPENTS.

The more we examine the works of creation, the more do we
behold with wonder the power of the Deity; and, having described some of the most powerful creatures of the tempes-
tuous ocean, I have now to call the reader's attention to a
few crude zoological observations on two species of animals
hitherto unknown to naturalists, and in all probability will
long remain so, together with many others; the existence of

which the comprehensive mind of man is unable even to conjecture.   We cannot, however, doubt but that they have
been created by the Almighty for some wise purposes, known
to him alone, and in viewing his productions we may well
exclaim in the language of the Royal Psalmist:—

> " Oh! Almighty God!
> How strange thy works! how great thy skill,
> And every land thy riches fill;
> Thy wisdom round the world we see,
> This spacious earth is full of thee:
> Nor less thy glories in the deep,
> Where fish in millions swim and creep;
> With wond'rous motions, swift or slow,
> Still wand'ring in the paths below.
> There dwells the huge *Leviathan*,*
> Who foams and sports in spite of man."   Dr. WATTS.†

SPECIES I.

THE AMERICAN SEA SERPENT.

THE animal I am now about describing, created a few years
ago many curious conjectures as to its actual existence, both
in England as well as in America; however, from the information I have gathered, I have no doubt of its being one of those
unknown animals which occasionally puzzle the zoologist
when they make their appearance.   I may observe that the
account I now present to the reader is derived from my friend
Sir Arthur de Capell Brooke's delightful record of his Travels
in Norway, &c.

---

* The term " *Leviathian*" has been supposed by Milton, and many
Scriptural authors, to signify the " *Whale;*" others suppose it to be the
" *Crocodile.*" But as alluded to in the Sacred Volume, I venture an
opinion that it is the subject of the article: at all events, it would not be
inappropriate, if applied to it.

† Vide Psalm civ. v. 18, 19, & 20, in Dr. Watts's Collection of
Psalms and Hymns.

This gentleman informed me, that when travelling through Norway, he made many enquiries after the Sea Serpent, but having been unable to see the creature himself, he was forced to be contented with the accounts he received. In 1818, the fishermen at Sejerstad stated, " That a sea serpent was seen in the Folden *fiord*, the length of which, as far as it was visible, was sixty feet. In July 1819, it made its appearance off Otersun in Norway, and Captain Schilderup stated to Sir Arthur, that it was seen daily during the whole of the month, and continued while the warm weather lasted, lying motionless as if dozing in the sunbeams. When Captain Schilderup first saw it, he was in a boat at the distance of about two hundred yards, and supposes its length to have been about three hundred ells, or six hundred feet. It was of considerable length, and longer than it appeared, as it lay in large coils above the water to the height of many feet. Its colour was greyish. He could not distinguish from the distance, whether it was covered or not with scales; but when it moved, it made a crackling noise, which he distinctly heard. Its head was shaped like a serpent, but he could not tell whether it had teeth or not. He said it emitted a very strong odour; and that the boatmen were afraid to approach near it, and looked on its coming as a bad sign, as the fish left the coast in consequence." Some fishermen at Forvig considered it possessed teeth, and to be about thirty feet long, and the head of a blackish colour. The bishop of Nordland and Finmark saw two in the bay of Shuresund, or Sorsund, in the Drontheim *fiord*, about eight Norway miles from Drontheim: he was not far from them, and considered the largest to be about one hundred feet in length.

In 1817, an animal having the form of a serpent, made its appearance in Gloucester harbour, and it was seen repeatedly by great numbers for some time afterwards, on different parts of the neighbouring coasts, particularly those contiguous to Cape Ann and Marble Head. The head and tail of this creature resembled that of the common snake, and generally lay in serpentine folds upon the surface. The size of the body was about that of a barrel, in colour nearly black, and of a hard and scaly appearance. The jaws were furnished with formidable teeth, and the whole length of the animal was supposed to be about one hundred feet. The folds were like wooden buoys floating: sometimes fourteen were visible, at others more or less. Those who saw it, agreed that it had neither gills nor fins. It moved like a caterpillar, but wonderfully rapid. The time that this animal remained in Gloucester harbour was about three weeks, and it was frequently seen pursuing the shoals of fish, which were observed to be unusually numerous, and were supposed to have occasioned its appearance.

During the first part of its visit, it seemed fearless of any thing, remaining stationary for some time, and suffering boats to approach near it. Numerous attempts, however, were soon made to take it; and the offer of a large reward occasioned contrivances of all kinds to effect its capture, dead or alive. Harpoons, guns, and swivels, were directed against it to no purpose; and though the balls from the latter were said to have struck it, no effect was produced. Large hooks were baited for it, and strong nets were set in different parts of the harbour, in the hope that the animal, in pursuing the fish might be entangled in them. All these were unsuccessful; and the creature being harpooned, soon became so shy and wary, that no boat could get near it, and the swiftness with which it darted away prevented any aim being taken at it. It finally disappeared, and was unheard of until the middle of August 1819, when an animal of a similar nature was seen off Nahant in Boston, where it remained for some weeks on that part of the coast. The first time it was visible, it was stationary for four hours near the shore, and two hundred persons assembled to view it. Thirteen folds were counted, and the head, which was serpent-shaped, was elevated two feet above the surface. Its eye was remarkably brilliant and glistening: the water smooth, and the weather calm and serene.

When it disappeared, its motion was undulatory, making curves perpendicular to the surface of the water, and giving the appearance of a long moving string of corks.

The repeated accounts of the serpent's appearance having attracted the attention of the Linnæan Society at Boston, one of its members was deputed to visit the spot, and examine into the truth of them. This was accordingly done, and the above is the general substance of the various depositions sworn to before General Humphreys. This gentleman, who was a corresponding member of the Society, despatched to the late Sir Joseph Banks copies of the whole of these, which are still preserved in his library. Sir Joseph entered with warmth into this curious investigation; and the minuteness with which every particular was supplied him showed how greatly he felt interested in the question.

In 1822, one of these creatures, the size of a large ox in circumference, and about a fourth part of an English mile in length, made its appearance off the island of Soröe, near Finmark, and was seen by many of the islanders.*

SPECIES II.

THE OPHIOGNATHUS AMPULLACENS, OR
BOTTLE-SHAPED SEA SERPENT.

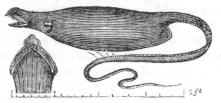

" A wondrous monster of the deep."

THIS new marine serpentiform animal is the subject of an interesting paper which was read before the Royal Society,

---

* THE SEA SERPENT RETURNED.—The Boston and New York Papers of July 9, 1833, state that the Sea Serpent has again appeared off Nahant:

about six years ago, by Dr. Harwood, the eminent Professor of Zoology, in the Royal Institution of Great Britain; and which from its partial resemblance to the Ophidian reptiles, and its large size, Dr. Harwood gave it the above name.

The specimen in question was unknown to zoologists, and was fortunately captured by Captain Sawyer, of the ship Harmony of Hull, in Davis's Straits, when he was in pursuit of the *Delphinus Deductor* (the bottle-nosed or Càaing whale). It was discovered lying on the surface of the water, and was at first supposed to be an inflated eel-skin, as employed by the Esquimaux to attach to their harpoons, for the purpose of wearing out the larger aquatic animals by its buoyant powers. From its continued endeavours to apparently gorge a species of perch, of greater circumference than itself, it was in a very exhausted state; and was easily taken. Captain Sawyer brought home the animal preserved in rum. It measured about four feet six inches in length, was very slender, and the tail has a filamentous termination, occupying about twenty inches of the entire length of the animal. The colour was a purplish black, the filamentous portion of the tail being lighter than the rest. From near the extremity of the snout, a sac extends about twenty inches down the body: this appendage, when partly inflated, is about nine inches in circumference, and its greatest width, including the slender body of the animal, is four inches; its only use appears to be that of a float. The animal has a single row of teeth above and below, no teeth on the palatal bones, and is destitute of a tongue. The jaws are so long, and their articulation of such a nature, that their opening is wider than even that of the rattlesnake. Of its habits, &c., nothing is known.*

He was first seen on Saturday afternoon, passing between Egg Rock and the Promontory, wending his away into Lynn Harbour, and again on Sunday morning, heading for South Shore. He was seen by forty or fifty persons (ladies and gentlemen), who insist that they could not have been deceived. *Extracted from the London " Times," August 1st.* 1833.

* I am indebted to the kindness of Mr. Limbird, the Publisher of the " MIRROR," for the engraving which illustrates this article.

CLASS.—PISCES, OR ARCTIC FISHES.

> " Fish that through the wet
> Sea-paths in shoals do slide, and know no dearth.
> O JEHOVAH, our Lord, how wondrous great
> And glorious is thy name through all the earth!
>
> MILTON.*

ORDER.—APODES.

GENUS.—XYPHIAS.

SPECIES.—XYPHIAS GLADIUS COMMUNIS,

OR

THE COMMON SWORD-FISH.

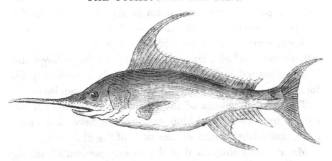

THE common sword-fish is a native of the Mediterranean and Sicilian Seas; it grows to a very large size, sometimes measuring twenty feet in length; it is active and predacious,

---

† Translation of the Eighth Psalm in the Magnet edition of Milton's Poetical Works, with a Biographical Sketch of the Author, by H. W Dewhurst, p. 506, recently published.

feeding on all kinds of fishes, and is likewise a formidable
enemy to the whale,* which it destroys by piercing with
its sword-shaped snout. Its body is long, round, and
gradually tapering towards the tail; the head is flattish,
but the mouth, and both jaws ending in a point, the upper
being stretched to a great distance beyond the lower; this
part, which is commonly called the "SWORD," is flattish
above and beneath; it is sharp on the sides; it is of a bony
substance, covered by a strong skin or epidermis; down the
middle of the upper part runs a furrow or impressed line,
and three similar ones on the lower surface: the tongue is
free and unconnected with the palate, and is of a strong
texture; in the throat are certain rough bones; the nostrils
are double, and seated near the eyes, which are moderately
large and protuberant. The body is covered by a thin skin,
having a fatty membrane beneath; the lateral line is placed
near the back, and is formed of a series of longish black
specks: the dorsal fin is very high at its commencement,
and sinking suddenly, becomes very shallow; it is continued
to within a small distance from the tail, terminated in an
elevated process: the ventral fin is placed nearly opposite
the part beneath, is moderately small, and much wider at
each extremity than at its middle. The pectoral fins are
rather small, and of a lanceolate form; the tail is large and
crescent-shaped, and on each side the body immediately before the tail is a strong finny prominence or appendage.

The several colours of the sword-fish is brown, accompanied by a deep steel-blue cast on the head and upper
parts, and silvery white on the sides of the abdomen.

Mr. Pennant observes that the ancient method of taking
the sword-fish, described by Strabo, agrees exactly with
that practised by the moderns of the present day. A man
ascends one of the cliffs that overhang the sea, and as soon
as he discovers the fish gives notice, either by his voice or

* See page 89, &c.

by signs to his colleagues, of the course it takes. Another
person, stationed in a boat, climbs up the mast, and on seeing
the fish, directs the rowers to it. As soon as he thinks they
have got within reach, he descends, and, taking a harpoon
or spear in his hand, strikes it into the fish, which after
having wearied itself with its agitations is soon killed; it is
seized and drawn into the boat. The flesh is much esteemed
by the Sicilians, who cut it into pieces and salt it: this process was anciently performed at the town of *Thurii*, in the
bay of *Tarentum*, and hence the fish was called Tornus
Thurianus.*

The sword-fish is found in the Northern Seas, and occasionally in the Pacific; it is probable, however, that it
has been often confounded with one of a different species,
and which is more common in that ocean.†

---

* Plin. lib. 32, c. ii.
† Shaw's Zoology, vol. iv.   Part i. p. 101.

ORDER.—JUGULARES.

GENUS.—GADUS.

SPECIES I.

GADUS MORRHUA,

OR

THE COMMON COD-FISH.

THE common Cod-fish* is a genus of the order Jugulares. The generic character is, pectoral fins slender and tapering to a point. The body is long, thick, laterally compressed, and covered with small smooth scales, which easily rub off. The head is smooth, wedge-shaped, with a broad front. The mouth large, the jaws armed with little sharp teeth bending inwards; some species have barbles hanging from the lower jaw. The tongue is broad and smooth; but the palate is rough, being armed with small teeth; and there are several rugged bones about the throat. The eyes are near the top of the head, round, large, and covered with a membrane. The nostrils are double and near the eyes. The coverts and opening of the gills are large; the covert consists of three laminæ, the under one edged with a skin. The membranes of the gills are strong, with seven or eight rays. The fins are from seven to ten, two pectorals, the

---

* SYNONYMES.—*Ash-coloured Cod.*   *Common Cod Fish*, Willoughby, Ray, Pennant, Shaw, and Donovan. *Gadus Morhua*, Linnæus. *Morhua*, Bellon. *Molva* vel *Morhua*, Rondelét. *Cablia*, Ström. *Gadus Squamus Majoribus*, Bloch.

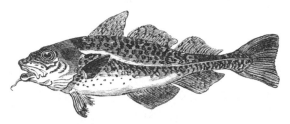

The Gadus Morrhua.—v. 214.

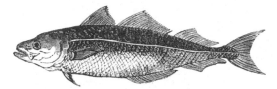

The Gadus Carbonarius.—p. 221.

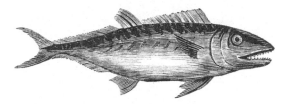

The Scomber Scombrus.—p. 224.

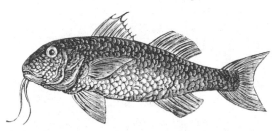

The Mullus Barbatus.—p. 227.

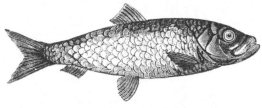

The Clupea Harengis.—p. 233

same number at the throat ; sometimes two at the anus, the tail-fin, and one, two, or three, on the back ; the rays of the fins are mostly covered with the common skin. The differ-ent species are found in the North and Baltic Seas, and some also in the Mediterranean and Western Ocean. They are all sea fish, except the Barbot, and do not often come into rivers.

The common Cod-fish inhabits the ocean, and is found between the fortieth and sixty-sixth degrees of North lati-tude ; it is also found in higher latitudes, as in Greenland, but then they are not so good nor so numerous. They are taken in vast quantities at Newfoundland, Cape Breton, Nova Scotia, and on the coast of Norway and Iceland ; also on the Dogger Bank, and about the Orcades. But their principal resort for centuries past has been on the banks of Newfoundland, and other sand-banks off Cape Breton. That extensive flat seems to be the broad top of a subaque-ous mountain every where surrounded with a deeper sea. Hither the Cod annually repair in numbers beyond the power of calculation, to feed upon the worms that swarm upon the sandy bottom. Here they are taken in such quan-tities as to supply all Europe with a considerable quantity of provision. The English have stages erected all along the shore, for salting and drying them ; and the fishermen, who take them with the hook and line, draw them as fast as they can throw out. This immense capture makes no sensible di-minution of their numbers ; for after their food is consumed in these parts, or when the season of propagation approaches, they take their departure for the Polar Seas, where they de-posit their roes in full security, and repair the waste which has been occasioned by death, or the depredations of their enemies. They annually make their appearance on the coast of Iceland, Norway, and Britain, gradually diminishing in their numbers as they proceed to the south, and ceasing altogether before they advance to the Straits of Gibraltar. Before the discovery of Newfoundland, the greatest fisheries

of the Cod were on the coasts of Iceland and the Western
Isles of Scotland, where the English resorted in quest of
them as early as the beginning of the fifteenth century.
Our right of fishing on these parts, however, was not acknowledged by the government of Denmark, till the reign
of James I., whose marriage with a princess of that country
secured to his subjects the indulgence, of which they availed
themselves so completely that they had then one hundred
and fifty ships employed in the Iceland fishery. Even on
the banks of Newfoundland, the French, Spaniards, and
Portuguese, had originally a far larger portion of the
fishery than the British; in 1570, the former nations had
upwards of three hundred vessels employed in that trade,
when those of the English did not exceed fifty. Matters,
however, have since been reversed; and the English shipping on that coast has immensely increased: it is now superior to that of any other nation, and the trade is deemed
a valuable accession to the wealth of individuals, as well as
to the naval power of the empire. Twenty thousand British
seamen are at present employed in this fishery, which is conducted in a tract of the sea agitated by a perpetual swell,
and involved in continual darkness by means of a thick fog
that constantly hangs over it. In the Cod fish, the sight is
probably very imperfect; for almost every small body that
is agitated by the water attracts their rapacity, stones and
pebbles not excepted, for these are often found in their
stomachs. The general weight of the cod-fish on the
British coasts is from fourteen to forty pounds; some indeed
have been caught near eighty, but those of the middle size
are most esteemed for the table. Their time of spawning
is from January to April, when they deposit their eggs in
rough rocky ground. After having been exonerated of a
load containing frequently three millions of young, the
parent recovers its plumpness sooner almost than any other
fish; and is caught in good condition during almost the
whole summer. Schoneved remarks a kind appointment of

Providence in the immense fecundity of this fish; and in
that abundant supply which it affords to the inhabitants of
those bleak and frozen climes that are unfit for the production of grain. "The ichthyophagi of these barren regions,"
says he, "not only furnish themselves with a substitute for
bread, by drying this fish, but send a vast quantity of their
surplus stores to the supply of other nations." The numbers
and fertility of these fish seem indeed amply to justify the
grateful exultations of this writer; for they are such as will
for ever baffle all the efforts of man and the voracity of the
inhabitants of the ocean to exterminate their race, at least
while they are caught only with hook and line. Leeuenhoek
reckons the number of eggs in a middle sized cod to be
9,344,000. Brindley reckons them at four millions only;
but even this is sufficient to supply all that can be destroyed by fishery, if we consider what quantities spawn
every year.

The mode of preparing the cod-fish for preservation consists partly in drying it in the air, partly in salting it, or in
both. The first makes what is called in the North stockfish,
or dried cod; the second Laberdom, or salted green cod; a
third preparation is called Klipp fisch (rock fish), or white
cod. The Icelanders, who get almost their whole food from
this fish, take the greatest pains to prevent want by preserving it when it is plenty; they dry it, giving it then the
general name of stockfish; but there are two preparations,
one called stockfish split cod, and the other hangefisch. They
are prepared in the following manner: when the men have
landed their cargo, the women cut off the heads of the fish,
open the belly, draw out the stomach, &c., then split the back
withinside, and take out the back-bone except the three last
vertebræ; then they boil the heads and eat them fresh, and
the men take the gills for bait. They dry the bones, which
serve to make fires, or as food for their cattle; and they
make oil from the livers. The fish thus cut up, they spread
them on rocks till the wind has thoroughly dried them, which

is generally in three weeks or a month ; but, if a sharp wind
blows from the north, this is generally accomplished in three
or four days. When there are no rocks and the soil is
sandy, they make a bed of stones laid close together, turning the fish downwards upon them, that the inside may be
kept from the rain, which would spoil it; then they suffer
them to lie in heaps till they find occasion to sell them. The
hangefisch is prepared nearly in the same manner; but with
this difference, that the back is cut from without, and split
entirely through, on which they are hung on stages of stone ;
as these stones are only laid one upon another without any
cement, the air has a free passage ; and the whole is
covered with boards or grass to keep out the wet.

The curing of this fish is different among the Norwegians
from what it is with the Icelanders, as they use salt. When
the heads are cut off and the fish cleaned out, they are put
into a large tub with a quantity of French salt ; and a week
after they put them into heaps on a kind of grating, to let
the blood and brine run off; after this they rub them with
Spanish salt, then pack them up tight in casks for sale,
under the name of Laberdom ; when only dried on the rocks,
they call them " Klipp fische." They split the large ones that
the salt may penetrate better, but the small ones have only
the belly opened; the latter are called round fish, the
former flat fish : when dried on poles, they call them "coth
fisch." All these sorts are carried to Bergen, and thence
transported to all parts of Europe. The heads are eaten,
but when forage is scarce they are given to cattle. The
inhabitants of the north dry the heads on the shore, and give
them to their cattle mixed with sea-weed : cows fed in this
manner we are assured give much more milk than when fed
on straw or hay.

As the air-bladder of this fish is very glutinous, the Icelanders prepare from it an isinglass not much inferior to that
of Russia. The following is the mode they pursue : they
leave the back-bones in lumps with the air-bladders attached

to them till they are ready to rot; then they lay them on a block, and heat the vertebræ till the bladders are loosened, as well as the ligaments by which they were fastened, and which are called pockets; they then cut away the bladders, and place them on a table or block, on which is nailed a rough brush intended to clean them; with a jagged knife they scratch the outer skin off the bladders and ligaments. The bladders being thus cleaned, they are soaked for a time in lime-water, to take off all the salt which may yet adhere to them; then they rince them in clear water, afterwards dry them, and then they may be used as isinglass. At Newfoundland they have attempted the same thing; but, as they have neither time nor room for all these processes, they salt the air-bladders and keep them for use, or else eat them. When wanted for the purpose of making isinglass, they must be soaked in water to take away the salt. The thickest bladders are the best for use, though the isinglass will not be quite so clear as that made from thin ones. The Norwegians eat the air-bladders fresh, or salt them for sale; they call them sunde mauer, or stomachic, believing them to be good for the stomach: hence the English name sounds or zounds. At Newfoundland they also turn the tongues to advantage, eating them either fresh or salted, and count them a delicacy. The Norwegians, Icelanders, &c., make oil from the livers, for when it attains a certain degree of corruption the oil of itself will run out, which oil is preferred to whale oil, because it keeps leather a longer time moist, and when used for light makes less smoke. The ova or eggs are also preserved with care, being salted and put into casks, and sold to the Dutch and French, who use them for a bait to catch anchovy; they export from Bergen annually at least twenty thousand barrels of these eggs; each barrel fetching about seven shillings sterling.

Though this fishery is very considerable in the north of Europe, it is not to be compared with that carried on in North America, particularly on the Banks of Newfoundland, as alluded to in the early part of this description. We may suppose that the Newfoundland fishery is not so considerable for the English, as before the rupture with the American colonies, as they now share it with us; the fishermen of Boston only taking annually in Massachusetts Bay little more than fifty thousand quintals of cod-fish, and a certain space in North America has also been ceded to the French for the like purpose. In 1785 the fishery was very plentiful for Great Britain, when two hundred and twenty vessels were employed, Pennant assuring us in his Arctic Zoology, that such a quantity of the cheven or chub fish are caught by us off the Dogger-bank and the Wall-bank, as to make us amends for the diminution of the fishery at Newfoundland.

According to Frazier, the cod appears in great numbers at Chili in the month of November. Admiral Anson says they are prodigiously large, and from the testimony of some of his people who had been at Newfoundland, they assert that they are as plenty there as at Newfoundland itself. Abbé Molina asserts, that the cod fishery is so plentiful on the coasts of Gio-Fernandez, that the same thing may be observed of it as we have remarked on Newfoundland, viz. that fish are drawn as fast as the lines are thrown out.* The cod appears also on the coasts of Valparaiso in the months of October, November, and December, but only in stormy weather; and these people, who formerly set no value on the cod-fish, have within these few years applied themselves to this lucrative branch of trade, and dry annually a great number of these fish; a Frenchman first, of the name of Luison, set them the example in their country. The cod is not tenacious of life: it dies soon after it is taken out of the sea, or even if put into fresh water. As they are excellent

---

* In 1824, on our return to England from the Arctic Regions, we caught one Sunday afternoon upwards of sixty cod-fishes in about half an hour, off Iceland.

when eaten fresh, the Dutch fishermen endeavour, by means
of perforated tubs, to convey them alive to the sea ports.
But the English prick the air-bladder with a pin, which
obliges them to remain at the bottom of the vessel they are
put in, and this preserves them longer alive.   The stomach
is large, and at the origin of the intestinal canal there are
six appendages divided into several branches.   The liver is
of a pale red colour, consisting of three lobes ; the spleen is
blackish and long; the kidneys are at the back-bone, along
the ventral cavity; the seed-vessel and ovary are double.

The cod generally haunts deep places in the open sea,
coming on the banks and shores in spawning-time ; they eat
crabs, whelks, herrings, and other fish, and are so greedy
that they do not spare even their own young ; like birds of
prey they have the faculty of casting up what they do not
digest, but according to Anderson, their digestive powers
are so strong, that the fishermen of Heligoland have found
the haddocks which were thrown out for baits completely
digested in their stomachs in six hours.

SPECIES II.

THE GADUS CARBONARIUS,

OR

COD COAL FISH.*

THE Gadus Carbonarius is an inhabitant of many of the
rocky coasts in Great Britain, more especially towards the

---

* The Cod Coal Fish, or Coal Cod, takes it name from the black
colour which it sometimes assumes.   Belon calls it the *Colfisch*, imagin-
ing that it was so named by the English zoologists, from its producing the
Ichthyocolla or *Isinglass*, but Gesner has given the true etymology."—
Sir A. Brooke.

northern parts of the island, as it is also throughout the
North and Baltic Seas. Captain Phipps procured this fish,
which, together with the Unctuous Lump-sucker, formed
the whole produce of his trawling and fishing, in animals of this kind, during his stay in the vicinity of
Spitzbergen. A small species of Gadus, nearly allied to
this, was found by Captain Scoresby among the Arctic
Ice, in the parallel of 78º N. Mr. W. Swainson, who
examined this specimen, considered it as a variety of the
coal fish; although, from the contraction and change of
colour produced by the spirit in which it was preserved,
the lateral line so essential in the determination of the species
could not be traced. The number of rays in each of the
fins of this was as follows: first dorsal thirteen, second dorsal eighteen, third dorsal twenty; first anal twenty-two,
second anal twenty, &c. The third ray of the ventral fin
was lengthened into a sibulate point; and the hinder dorsal
fin rounded: these are peculiarities that have not been
noticed by another, who has hitherto described the coal-
fish.

According to Gmelin, this fish is likewise found in the
Pacific Ocean. According to Mr. Pennant, the young of
this species swarm about the Orkney Islands and the Coast
of Yorkshire; in the former they constitute the chief support of the poor, and the same writer adds that the fry is
known by different names in different places; at Scarborough,
for example, they are called *pans*, and when a year old
billets. About 1802-3 such numbers of these young animals
visited that part that, for several weeks, it was impossible
to dip a pail into the sea, without taking some.

A late author states that although this information may
be correct, yet he cannot avoid expressing his opinion that
it is the fry of the Salmon, and not the coal-fish, to which
Mr. Pennant alludes, in detailing the history of the latter
species. Nothing at least, he observes, can be more evident than that the *Pan* of Mr Pennant, and to which he

refers as the young of the Gadus Carbonarius, is no other than the fry of the Salmon.*

Though this fish is so little esteemed when fresh, yet it is salted and dried for sale, one person having in a year cured above a thousand of the *true* coal-fish at Scarborough.

The fry of the coal-fish form a considerable portion of the food of the inhabitants connected with the coasts to which it resorts. About the commencement of July, the young fish appear on the coasts of Yorkshire, in vast shoals, and are at this time about an inch and a half long. In August they are from three to five inches in length, and are taken in great numbers with the angling rod, and are esteemed a very delicate fish, but when a year old the flesh is very coarse, so that very few persons will eat them. Fishes of that age are from eight to fifteen inches long, and begin to have a little blackness near the gills and on the back ; as the animal increases in age, this blackness increases.

The Linnæan character of this fish is accurate and concise. M. Bloch † conceives the blackness and straightness of the white lateral line as sufficient to discriminate the species. The skin of the coal-fish is covered with very small oblong scales. It has three dorsal fins, in the first of which are about fourteen rays, in the second eighteen, and in the third twenty; in the pectoral fin eight ; ventral fin five ; first anal fin twenty-six ; second twenty, and the tail thirty-three, besides ten very short ones on each side ; the tail is broad and of a forked shape.‡

The colour of this fish varies considerably in the different stages of its growth. When young, a certain duskiness prevails in the colours, which becomes darker as the fish grows older; and when they have attained their full size, which is about two or three feet, the back, nose, dorsal fins,

* Rees's Cyclopedia, vol. xv. Art. *Gadus.*

† Block, Ichthyologist, on Hist. Naturelle, générale et Particuliere des Poissons. Planche lxvi.

‡ Donovan's British Fishes.

and tail oftentimes appears of a deep black.   Beneath the
pectoral fins, a black spot is perceivable ; the mouth is of
the same colour, whilst the tongue assumes a bright silvery
hue.   The irides of the visual organs have a similar appearance, and marked on one side with a blackish spot.
The Gadus Carbonarius is of a much more elegant form
than the common cod-fish.   Its weight in general is not
more than twenty or thirty pounds.*

ORDER.—THORACICI.

GENUS.—SCOMBER.

SPECIES.—SCOMBER *SCOMBRUS*,

OR

THE COMMON MACKEREL.*

This very beautiful fish is a native of the American Seas,
generally appearing at stated seasons, and swarming, in vast
shoals, around particular coasts; but its great resort is within
the Arctic circle, where it resides in innumerable troops,
and grows to a larger size in those regions than elsewhere,
and consequently from this cause, we may naturally suppose
that it there finds its favourite food in greater quantities than
in the warmer latitudes, which is found to consist of chiefly

---

* Sir Arthur de Capell Brooke saw shoals of young coal cod-fishes in
the Norwegian Seas, and the sea was darkened by them to a great extent.
The Norwegians denominated it the " *Sey-fish.*"   Vide *Travels in Norway,* &c.

* Synonyme.—Scomber Communis.

marine insects. According to Admiral Le Pleville le Peley, they have been observed by him in severe winters, and even to the commencement of spring, about the coasts of Greenland and Hudson's Bay, at the bottom of the small clear hollows, encrusted with ice round the coasts, entirely bristled over with the tails of the mackerels, whose heads were imbedded in the fine soft mud at the bottom, to the extent of more than three inches, in some even to three parts their entire length, being thus protected sufficiently from the severe effects of the frost; and on the return of the spring, they are generally believed to migrate in enormous shoals, of many miles in length and breadth, to visit the coasts of more temperate climates, in order to deposit their spawn. Its route has been supposed to be similar to that of the Clupea Harengis or common herring, passing between Iceland and Norway, and proceeding towards the northern parts of Scotland, the Orkney and Shetland Isles, where a part of the shoal throws itself off into the Baltic, whilst the grand column passes downwards and enters the Mediterranean Sea, through the Straits of Gibraltar.

This long migration of the mackerel, as well as the herring, seems at present to be greatly called in question; and it is considered as more probable that the shoals which appear in such abundance round the more temperate European coasts in reality reside during the winter at no very great distance, immersing themselves in the soft bottom, and remaining in a state of torpidity, from which they are awakened by the genial warmth of the returning spring and gradually recover their former activity. At their first appearance their eyes are observed to appear remarkably dim, as if covered with a kind of film, which passes off as the season advances, when they appear in their full perfection of colour and vigour.

The general length of this species of the mackerel is from ten to sixteen inches, but in the Arctic Seas it is occasionally found of far greater size, and, amongst those which visit our

own coasts, instances sometimes occur of specimens far exceeding the general size of the rest. The colour of this
fish, on the upper parts, as far as the lateral line, is a very
deep blue, accompanied with a varying tinge of green, and
marked by numerous black transverse streaks, which in the
male are nearly straight, but in the female beautifully undu-
lated; the jaws, gill covers, and abdomen, are of a bright
silver colour, with a slight varying coat of gold-green along
the sides, which are in general well marked in the direction
of the lateral by a row of long dusky spots ; the scales are
very small, oval, and transparent; the finnules or spurious
fins are small, and are five in number both afore and below.* The shape of the mackerel is highly elegant, and it
is justly considered as one of the most beautiful of the European fishes. Its merit as an article of food is universally
established, and it is one of those fishes which have maintained their reputation through a long series of ages, having been highly esteemed by the ancients, who prepared
from it the particular condiment or sauce known to the
Romans by the name of *Garum*, and made by salting the
fish, and after a certain period straining the liquor from it.
This preparation, once so famous, has been long super-
seded by the introduction of the anchovy for similar purposes.

There are twenty-two species of the mackerel, but this is
the only one found in the Arctic Regions. The mackerel
is easily snared by a variety of baits; but the capture is said
to succeed best in a gentle gale of wind, hence the term a
" *Mackerel Gale.*"

---

* There is not a fish which exceeds the mackerel in the brilliancy of its
colours, or in the elegance of its shape. The fine deep blue upon its back
is crossed by many streaks, and accompanied by a tinge of green, which
varies as the fish changes its position. The bright silver colour of the
abdomen, and the varying tinge of gold-green which runs along the
sides, are eminently beautifully in this species ; but are only to be seen
to perfection when it is first taken out of the water, as death impairs the
colours.—*Wood's Zoography*, vol. ii. p. 170.

ORDER.—THORACICI.

GENUS.—MULLUS.

SPECIES.—MULLUS BARBATUS,

OR

THE COMMON RED SURMULLET.*

THE red Surmullet is principally found in the Mediterranean and Northern Seas, and one of Captain Scoresby's seamen took one out of the mouth of a seal near Spitzbergen: the body was about twelve inches; however its extreme length is generally about fifteen inches. Its colour is of an elegant rose red, tinged with olive-colour on the back, and of a silvery cast towards the abomen; the scales are thin, and easily rubbed off, the skin itself appearing of a brighter red. The surmullet is a fish of a strong and active nature, swimming briskly, and feeding principally on the smaller fishes, worms, and sea insects.

This fish was highly esteemed by the Romans, and bore in consequence an exceedingly high price. When boiled it constitutes an excellent dish. The capricious epicures of the time of Horace† valued it in proportion to its size ; not that the larger were more delicious, but more difficult to procure. The price that was given for one in the time of

---

* SYNONYME.—Mullus Ruber.   † Juvenal Sat. iv. 48*l*. 8*s.* 9*d.*

Q

Juvenal and Pliny is a very striking evidence of the luxury and extravagance of the age.

" Mullum sex millibus amit
*Æquantum sane paribus sestertia libra."** *

" The lavish slave
Six thousand pieces for a mullet gave,
A sesterce for each pound."

DRYDEN.

But Asinius Celer, a man of consular dignity, gave a still more unconscionable sum; for he did not scruple in bestowing eight thousand pieces, a sum equal to sixty-four pounds eleven shillings and eight pence of our money, for so small a fish as the mullet; for, according to Horace, a *Mullus trilibris,* or one of three pounds, was a great rarity, so that Juvenal's spark must have obtained a great bargain in comparison of what Celer had.

But Seneca informs us that it was not worth the amount of a farthing, except it had died in the very hand of the guest; and such was the luxury of the age that there were even stew holes in the dining-rooms, so that the fish could at once be brought from under the table and placed upon it. The Romans first brought the mullets in transparent glass vases to the table, in order that they might enjoy the felicity of contemplating the beautiful changes of its evanescent colours during the time of its gradual expiration, after which it was prepared for their repast.†

Apicius, a wonderful genius for luxurious inventions, first hit upon the method of suffocating them in the exquisite Carthaginian pickle, and afterwards procured a rich sauce from their livers.

The lips of the mullet are membraneous, and the lower one carinates inwards. They have no teeth in the jaws,‡ but

* Plin. lib. ix. c. 17.
† Seneca Nat. Quæt.   Lib. iii. c. 17.
‡ Bingley's Animal Biography, vol. 3. p. 292.

on the tongue and palate, above the angle of the mouth is a hard callus: the gill-membrane has seven incurvated rays. The gill-covers are smooth and rounded.

Mr. Pennant states that he heard of this species having been taken on the coast of Scotland, but had no opportunity of examining it; and whether it is found in the West of England, with the other species, as yet is unascertained: at least it is but rare. Salvianus makes it a distinct species, and says that it is of a purple colour, striped with golden lines, and that it did not commonly exceed a palm in length: it is no wonder that such a prodigy should captivate the fancy of the Roman epicure. Mr. Ray* establishes some other distinctions, such as the first dorsal fin having nine rays, and the colour of that fin, the tail and the pectoral fins, being of a very pale purple. However this is described by Mr. Pennant under the title of the striped Surmullet.

ORDER.—ABDOMINALES.

GENUS.—SALMO.

THE generic character, as given by Gmelin, of this genus of fishes, is as follows:—Head smooth and compressed, with a large mouth, small lips, tongue white, cartilaginous, and movable; eyes are moderate, lateral; teeth in the jaws and on the tongue; the gill-membrane is from four to twelve rayed, but the cover has three laminæ; the body is long, covered with rounded and very finely striate scales; the back is convex; the lateral line is straight, near the back; the hindmost dorsal fin is fleshy without rays; the ventral fins have moving rays.

---

* Synopsis Piscium.

SPECIES I.

SALMO GRÖENLANDICUS,

OR

THE GREENLAND SALMON.

THE length of this species is about seven inches, which it very rarely exceeds; shape lengthened, contracting somewhat suddenly towards the tail; the dorsal fin is placed in the middle of the back; fins rather large for the size of the fish; small scales, forked tail, colour pale green, with a tinge of brown above; abdomen and sides silvery; in the male fish just above the lateral is a rough fascia, beset with minute pyramidal scales, standing upright like the pile of a shag. The use of this villous line is highly singular, since it is affirmed that while the fish is swimming, and even when thrown ashore, two, three, or even as many as ten will adhere, as if glued together, by means of this pile, so that if one is captured the rest may be also taken at the same time. This species swarms off the coasts of Greenland, Iceland, Newfoundland, and Nova Scotia, and it is said to form part of the food of the Greenlanders. The inhabitants of Iceland dry it in great quantities, in order to reserve it as winter food to their cattle, whose flesh is apt to acquire an oily flavour in consequence. This fish lives in the sea the greatest part of the year; but in April, May, June, and July, they come into the bays in immense shoals, where great multitudes are taken in nets, and afterwards are dried on the rocks. When fresh they are by some zoologists said to have the smell of a cucumber, though others affirm the scent to be highly unpleasant. They feed upon small crabs and other marine insects, as well as on the smaller fuci and confervæ (*marine plants*), on which they are observed to deposit their ova.[*]

The following species of this genus are peculiar to the Northern Regions :—

2. SALMO SALAR, the common Salmon, is occasionally met with on the coasts of Greenland, and all over the North of Europe and Asia, from Great Britain to Kamtschatka.

3. SALMO APINUS, the Red Char, is found in the cold lakes of the Lapland Alps, where it is fed on the larvæ of gnats that infest those regions. The Laplanders in their migrations to the distant lakes during summer, find a ready and luxurious repast on these fishes, which to them are extremely palatable.

4. SALMO TAIMEN, the Brown Salmon, inhabits the rivers which empty themselves into the Frozen Ocean.

5. SALMO KUNDSCHA, inhabits the bays of the Arctic Seas.

6. SALMO ARCTICUS, the Arctic Salmon, inhabits the stony rivulets running into the Arctic Seas, it is not longer than one's finger, and resembles a young thymallus or Grayling. It is silvery, with four rows of brown dots, and fine lines on each side, the tail is forked.

7. SALMO STAGNALIS, inhabits the remoter mountainous rivers of Greenland.

8. SALMO RIVALIS, inhabits the muddy rivers and stagnant lakes of the Icy Regions.

9. SALMO AUTUMNALIS. This fish inhabits the Frozen Seas, and ascends in vast shoals the rivers which empty themselves into it, periodically, chiefly in the autumn.

10. SALMO PELED, the Peled Salmon, found about Nova Scotia and the Northern parts of Russia.

---

[*] I have not been able to procure a drawing of the Greenland Salmon on which I could rely for accuracy ; but, should I succeed, one shall be furnished gratuitously to the subscribers of this volume.

11. SALMO THYMALLUS, the Grayling Salmon, found about Lapland, the natives of which use its intestines in lieu of rennet in preparing the cheese which they make from the milk of the rein-deer.

GENUS.—CLUPEA.

SPECIES.—CLUPEA HARENGIS,

OR

THE COMMON HERRING.

OF the herring genus there are three species; the common herring (the *clupea harengis*), the pilchard (*clupea pilcardus*), and the shad (*clupea alosa*), of which the fry has been erroneously considered as the white-bait of the estuary of the river Thames and other places; this, however, has been proved incorrect by that talented zoological anatomist, William Yarrell, Esq. Although the common herring is the immediate subject of this article, yet I may observe that this fish and the pilchard greatly resemble each other in size, being about twelve inches long when fully grown; but there are some obvious characterestic distinctions between them, as well as in their appearance and habitations. Their colour is nearly the same, but the pilchard is more elevated in the back and rounder than the herring; it is also blunter in the muzzle and has larger scales. However, the most obvious distinction between them is the position of the dorsal fin. In the pilchard it is placed exactly over the centre of gravity, so that if the fish were suspended by it, the body hangs in an horizontal position. But in the herring it is further back than the centre of gravity, so that the head droops when the fish is lifted by it. The same distinction is found in the fry as well as in the full-grown fish. According to Mr. Mudie, the fry of the herring and pilchard are taken in great numbers, and known by the common name of sprats.* This, however, is incorrect, and displays very little of Mr. Mudie's research, and exhibits his glaring deficiency both in zootomy and zoology. For the sprat is distinguished by its full length being only four or five inches ; its body being much deeper than that of a young herring of equal magnitude ; and its dorsal or back fin being still more remote from the nose. A still more remarkable distinction between the sprat, the herring, and pilchard, appears in the belly ; that of the two first being quite smooth, whilst that of the last is very strongly serrated. Again, if we refer to its zoological anatomy, we discover a most distinguishing character, in the spine, inasmuch as the herring has sixty-six vertebræ, and the sprat but forty-eight. The period too when the latter fish visit our coasts is another peculiarity between this, the herring, and pilchard ; they continue with us in large shoals, when the others have generally returned to their hyperborean deeps.†

The distinguishing characters of the herring are,—that there are eight branchiostegous rays, the belly is extremely sharp, and frequently serrated.

Herrings differ greatly in size, but their usual length is from nine to twelve inches. The back and sides are green, varied with blue, the belly is silvery ; a scaly line runs along the belly from the head to the tail; the scales, which are very large and thin, fall easily off; the eyes are very fine and large ; the edges of the upper jaw and the tongue are very rough, but the mouth is destitute of teeth. The gill-covers are large and patulous, which occasions the im-

---

* British Naturalist, vol. i. p. 282.

† Martins' Dictionary of Natural History, vol. ii. Art. Sprat.

mediate death of the fish the moment it is taken out of its
native element, and hence the well-known proverb,—

"As dead as a herring."

The dorsal fin consists of seventeen rays; the two ventral
fins have nine, the pectoral seventeen, and the anal fourteen.
The tail is extremely forked; the lateral line is hid beneath
the sides, and the sides are compressed.

Herrings are found in the greatest abundance in the
higher northern latitudes. In those unnavigable seas where
the ark of man ne'er found a resting place, from their being
covered with ice for the greater part, and in some places
the whole year round; beneath these continents of ice these
fishes find a quiet and safe retreat from their numerous enemies; thither neither man, the finner (*balænóptera jubartes*),
nor the physeter macrocephalus, or great-headed cachalot,
dare to pursue them. The quantity of insect food which
those seas supply is amazing (as I have already mentioned),
which, added to the security of these fishes beneath the icy
rigour of the climate, render their increase beyond all
calculation, and from these retreats some zoologists have
supposed they would never depart, did not their numbers
render it necessary for them to migrate in quest of food,
more congenial to their climate. The great colony of
herrings sets out from the Arctic Ocean about the middle
of winter, it is supposed in such quantities as defy conception; but no sooner is it in motion than millions of enemies
are on the alert, and thin their squadrons. The *finner*
and the cachalot swallow hundreds at one mouthful—the
porpoise, the grampus, shark, and the numerous tribes of
dog-fish, find them an easy prey;—and, desisting from their
carnage, unite in devouring these defenceless animals:
while the myriads of sea-fowl that inhabit and frequent the
Arctic regions watch the progress of their migration, and
thus form almost as fatal a class of enemies as those of the
vast and mighty deep.

Thus surrounded by foes, which they can neither avoid nor repel, these helpless emigrants find no other safety but in crowding close together, and leaving their extreme ranks to be first destroyed. However, they soon separate into shoals, one body of which moves westward, and pours down along the coasts of America, as far as Carolina. In Chesapeake Bay, the annual inundation of these fishes is so great that they cover the shores in such vast quantities as to become almost a nuisance. Those which hold more to the eastward, and direct their course for Europe, endeavour to save themselves from their unrelenting pursuers by approaching the first shore they can find; and accordingly make their descent upon Ireland about the commencement of March. When they arrive on that coast their phalanx, though it has suffered considerable diminutions, is nevertheless of amazing extent, depth, and closeness, covering a space as large as the island itself; the whole element seems as it were alive, and the numbers appear inexhaustible.

The shoal that visits the British coasts begins to appear off the Shetland Isles in the month of April. This is the forerunner of the grand shoal which descends in June, whose appearance is announced by the numbers of its voracious attendants, the gannet, the shark, and the porpoise. When the main body arrives, its breadth and depth is such as to alter the very appearance of the ocean. It is divided into distinct columns of five or six miles in length, and three or four in breadth; while the water curls up as the herrings advance, appearing as if forced from its bed. Sometimes they sink for the space of ten or fifteen minutes, then rising again to the surface, and in bright weather reflecting a variety of the most splendid colours, like a field bespangled with azure, gold, and purple.

From the Shetland Isles, where this great army divides, one body moves off to the western coasts of Ireland, where it meets with a necessity of dividing a second time : one party, taking to the Atlantic, is soon lost in that extensive ocean; the other, passing into the Irish Sea, furnishes a very considerable capture to the natives. The second grand division takes place at Shetland, visiting the northern shores of this island, and then entering the British Channel, passes the Land's End, in Cornwall, and soon after totally disappears.

Thus the herrings expelled from their native seas seek those bays and shores where food presents itself in the greatest plenty, and where they are the least liable to meet with their ferocious persecutors and pursuers of the ocean. In general, the larger rapacious animals of the sea avoid coming into contact with the more populous shores; hence we find that an unerring and all-gracious Providence has given a species of natural instinct to these species to approach these parts for a twofold purpose—the one in affording them a safe retreat from their enemies, and the other that they may be captured by man, and thus become subservient to his natural wants. Thus, all along the coasts of Great Britain, Norway, Germany, and France, they are found pretty punctual in their periodical visitations; nevertheless, they are sometimes capricious in their migrations, and have even been known to frequent particular shores for a series of years, and then to relinquish them for ever.

The statement I have above given, and which has been confirmed by every zoologist and ichthyologist whose works I have been able to peruse, is impugned by a Mr. Mudie, in his *British Naturalist* *—for no philosophical reason whatever, and, what is worse, he is unable to produce a single illustration whatever in corroboration of his objection; besides the learned gentleman from his *pursuits* is unable to form a correct idea of the place from whence these fishes owe their origin, and I am much rather disposed to believe the remarks of both Greenland captains and seamen who

---

* A work which, though amusing, unfortunately abounds with numerous errors.

have caught them, combined with the researches of zoologists, to that of an industrious author, whose sole observations are made in his private study, and are either the produce of his own fertile imagination or the hypothesis of some writer whose name Mr. Mudie has kept a secret. What Mr. M. has stated may *sometimes occur*, but it is incorrect to consider it as a general principle. The following are his observations on this subject: " The herring," he says, " come to the shores and estuaries (of the warmer latitudes) in order to mature and propagate their spawn, which they do over a greater range of the year than most other fish, continuing the operation to the middle of winter, and retiring into deeper water after that is done. But there is no reason to conclude that they have much migration in latitude, or that they ever move far from those shores which they frequent in the season. The fry too are found on the shores, and in the bays and estuaries frequented by their parents; and they do not go into the deep water until late in the season. They even appear to go farther up the rivers than the old fish, for they may be taken in brackish water, with a common trout-fly."

Towards the end of June, herrings are in full *roe*, continuing in full perfection until the commencement of winter. In fact, there is an old adage, " *That the roes disappear after the ninth of November*," the day on which the Lord Mayor of London enters on his duties as chief magistrate. This assertion is, however, incorrect, as I have, and in all probability many thousands besides, eaten them so late as December. The young herrings approach the shores in the months of July and August, and are then from half an inch to two inches in length. Few young herrings being discovered in our seas during winter, it is generally supposed that they return to their native haunts beneath the ice, in order to repair the vast destruction of their race in the summer. Some old ones continue on our coasts the whole year, but the number is very inconsiderable.

The herring fishery is of very remote antiquity. The Dutch, remarkable for their persevering industry, first engaged in it about the year 1164: they kept possession of it for several centuries; but at length its value became so well known that it gave origin to several obstinate contests between them and the English. Still, however, either from some defect in the British government, or an inferior method of conducting our fisheries, the Hollanders maintain a decided superiority over us in this lucrative branch of commerce.

Our great stations are off the Shetland and Western Isles, and on the Norfolk coast, in which the Dutch also participate. Yarmouth has long been famous for its herring-fair, which was regulated by an especial act of parliament in the reign of king Edward the Third. The town is annually compelled by its charter to send to the sheriffs of Norwich one hundred herrings, to be made into twenty-four pies, by them to be delivered to the lord of the manor of East Carleton, who is to convey them to the king; and hence the facetious Dr. Fuller denominated a herring, " a Norfolk capon."

Immense quantities of herrings are annually caught on the British coasts, many of which are consumed while fresh; and the rest are either salted, or smoke-dried, and exported to various parts of Europe.

When salt was subject to a high duty, and sufficient salt was not kept at those places where herrings make their capricious appearances, great loss was frequently sustained. This happened occasionally on many parts of the Scotch coast, but more particularly on the north of the entrance of the Frith of Forth. That Frith, as it is deep water, and without any shallow or interruption, is a favourable resort of herrings in the autumn and early part of winter. They come from the deep water in immense shoals or masses. For this reason they generally prefer deep water, avoiding the shoal coasts; and, when they do get entangled upon one, great numbers are wrecked.

The rocky promontory at the east end of the county of
Fife, off which there lies an extensive reef or rock, sometimes has that effect; and there have been seas in which,
when the difficulties of the place have been augmented by
a strong wind from the south coast, that carried breakers
upon the reef, and a heavy surf along the shore, the beach
for many miles has been covered with a bank of herrings
several feet in depth, which, if taken and salted when first
left by the tide, would have been worth many thousands of
pounds; but which, as there was not a sufficient supply of
salt in the neighbourhood, were allowed to remain putrefy-
ing upon the beach, until the farmers found leisure to cart
them away as manure. One of these strandings took place
in and around the harbour of the small town of Crail, only
a few years ago, but prior to the new regulations which were
enacted with regard to salt. The water appeared at first so
full of herrings that half a dozen could be taken at each
dip of a basket. Numbers of people thronged to the water's edge, and fished with great success; and the public
crier was sent through the town to proclaim that " callar
herrings," which signifies herrings fresh from the sea, might
be purchased at the rate of forty a penny. As the water
rose, so the fish accumulated, until numbers were stunned,
and the rising tide was bordered with fish, with which
baskets could be filled in an instant. The crier was upon
this instructed to alter his note, and the people were invited
to repair to the shore, and obtain herrings at a shilling the
cart-load. But every succeeding wave added to the mass
of fish, and brought it nearer to the land, which caused a
fresh invitation to whoever might feel inclined to come and
take what herrings they chose, *gratis*. The fish still continued to accumulate till the height of the flood, and, when
the water began to ebb, they remained on the beach. It
was rather early in the season, so that warm weather might
be expected, and the effluvia of so many putrid fish might
occasion disease; therefore the corporation offered a reward

of one shilling per cart-load to every one who would re-
move that quantity from the part of the shore under their
jurisdiction. The fish, being immediately from the deep
water, were in the highest condition, and barely dead. All
the salt in the town and neighbourhood was instantly put
in requisition, but it did not suffice for a thousandth part of
the mass, a great proportion of which, notwithstanding
some very successful attempts to carry off a few sloop
loads in bulk, was lost. In the bays or " *locks,*" on the
west coast of Scotland, where the shoals of herrings are
very abundant, and apt to be driven ashore and stranded
by heavy gales from the north-west, these casualties occur
frequently. But, though these events are attended with a
great and obvious loss, they do not appear to have any sensible effect upon the supply of herrings, whose numbers do
not seem capable of apparent diminution, either by the casualties of nature or the schemes of art.

The habits of this most abundant, and perhaps, all things
considered, most valuable fish, are but imperfectly known ;
and they have been greatly misrepresented.

Considered as an aliment, fresh herrings are perfectly innocent if but moderately used; they contain, however, a
great quantity of oil, which disagrees with weak and delicate stomachs; and, according to Dr. Martyn, if they are
eaten in disproportionate quantities, they disorder the digestive powers, frequently producing putrefaction in the digestive cavities, of an alkaline nature, and consequently are
attended with pernicious effects. Pickled and dried salt
herrings are always unwholesome food, and, although they
form what is commonly denominated " *a relish,*" yet their
flesh is hard, and difficult of digestion by the vital powers :
the former are, however, less injurious than the latter.

The ancient dietetians viewed herrings of great importance in a medicinal point of view. The vesicles, termed
animæ, were taken internally as a diuretic. Salted herrings
were applied to the soles of the feet in fevers, in order to

draw the humours from the head, and to mitigate febrile heat. Herring pickle was employed as a *lavement* in dropsy and pains in the hips—and externally to purify fœtid ulcers, to dissipate scrofulous swellings, and to arrest the progress of a gangrene. It is also said to be beneficial in *cynanche tonsilaris*, or quinsey of the throat, if the parts affected are anointed with a mixture of this and honey.

The Dutch herring-fishery commences on the 14th of June, in which no less than a thousand vessels are employed; these, which are called busses, carry from forty-five to fifty tons, besides two or three small cannon. None of them are allowed to quit their posts without a convoy, unless they carry twenty pieces of cannon collectively, in which case they are permitted to sail in company. Before they proceed on their voyages the owners make a verbal agreement, which carries in it all the force and authority of the most solemn compact. The regulations of the admiralty of Holland, with a few variations, are followed by the French and other nations: the principal of which are, that no fisher shall cast his net within one hundred fathoms of another's boat; that while the nets are cast a light shall be kept on the stern of each vessel; that when a boat is obliged by any accident to desist from fishing the light shall be cast into the sea; and likewise that when the majority of the fleet leaves off fishing the rest shall be obliged to do the same.*

---

* In Panamaquoddy Bay and the neighbourhood vast quantities of herrings are taken by scooping them up with hand-nets. The fishing is carried on during very dark nights, and often displays the most striking and picturesque appearance to the spectator ashore. The fishermen go in small light boats, each bearing a flaming torch. The boats row with great swiftness through the water, and the herrings, attracted by the glare of the light, crowd after the boats in such numbers that those stationed in the stern, for this purpose, scoop them up by thousands. The fish frequently throng together with so much eagerness as to throw one another out of the water.—GOODRICH.

From the middle of September to the middle of October is the most successful period for fishing on the Norfolk and Suffolk coasts. The nets which are used for herrings are about five yards deep, and twenty-five feet long; and sometimes such numbers of them are united that they will take in a mile in compass. The fishermen are directed to those spots where the herrings are most numerous by the hoverings and motions of the sea-birds, which continually pursue them in expectation of prey. As the fishermen row gently along, they let their nets fall into the sea, steering their course as nearly as they are able against the tide; so that, when they draw them, they may have the assistance of the tide. As soon as any boat has procured a lading, it makes to the shore, and delivers the fish to those persons who are appointed to wash and gut them. They distinguish their herrings into six different sorts: the fat herring, which is the largest and thickest of all, and will keep about two or three months; the meat herring, which is likewise large, but less fat and thick than the former; the night herring, which is of a middling size; the pluck, which has been somewhat damaged in catching; the shotten herring, which has lost its spawn, or milt; and the copshen, which by some accident or other has been deprived of its head.

All these kinds of herrings are deposited in a tub with salt or brine, where they are permitted to lie for twenty-four hours; these are then taken out, put into wicker baskets, and washed; after which they are fixed on small wooden spits, and hung up in chimneys built for that purpose, at such distances that the smoke may have free access to them all. When these places, which will contain ten or twelve thousand fish, are filled, a quantity of billets is laid on the floor, and set fire to, in order to dry them; and, the doors and air-holes being closely shut, the whole place is immediately filled with smoke. This operation is repeated every quarter of an hour, so that a single barrel of herrings requires five hundred billets to dry them. A last consists of ten barrels, and each barrel contains about a thousand herrings, which, when thus prepared and dried, receive the appellation of red-herrings.

The Dutch are most expert in pickling these fish, and for that purpose they take them about the middle of summer. Their usual method of procedure is as follows: as soon as the herrings are liberated from the nets, they are gutted and washed; then they are put into strong brine, made of water and sea-salt, for fifteen hours; after which they are taken out, well drained, and regularly disposed into barrels, with a layer of salt at the bottom of each, and another at the top. Care is likewise taken that no air be admitted, nor the brine suffered to leak, either of which would be injurious to the preservation of the fish.

The following article illustrates to a certain extent Mr. Mudie's theory, which I have had occasion to condemn; however, as the talented author has stated a fact, it is but right to give Mr. Mudie the benefit of it—in many other points, however, it is highly interesting to the general and scientific reader, therefore I have much pleasure in inserting it.

Observations connected with the migration of the herring and mackerel, as noticed in the British Channel, by Major W. M. Morrison.

HASTINGS, from its peculiar situation, is well situated for a fishing station, and has in consequence for a considerable period employed many vessels in this particular branch of commerce; each vessel is furnished with from one hundred to one hundred and twenty nets, each net being forty feet in length. They can be joined to each other with great facility, and, when in the sea, present a curtain from fourteen to sixteen feet in depth. These the fishermen, when at any distance from the land, always shoot or place north and south, or as near that direction as can be done conveniently, in order that they may drift with the flowing and ebbing of the tide, which takes the direction of east and west in this part of the British Channel. I have particularly noticed this latter circumstance, both attached to the capture of the herring and the mackerel, which is, that those fish encumbered with roes, while caught in great numbers on the east side of the nets, are not met with in a greater proportion than one in about one hundred without roes on the west side, a fact which affords evidence that not only the herring but also the mackerel reach this part of the channel, for the purpose of depositing their roes from the eastward.

When the nets are arranged for the mackerel, the upper part are always supported on the surface by small kegs and corks; but, when placed for the taking of herrings, they are not always left near the surface, but are sunk at various depths when there is little or no wind, from within a yard of the bottom upwards, according to the judgment of the fishermen, and they generally prefer placing them near the surface when there is a brisk breeze.

About November, herrings generally appear off Hastings, sometimes earlier: if, for instance, the wind sets in from North-west in the beginning or middle of October, occasioning naturally smooth water along the east coast of England, then the advance of the herrings southward is greatly facilitated. Should this continue for some time it ensures a profitable season to the fishermen of this place. Should a south or south-east wind come on and prevail for some time while the herrings are on their passage to the channel, it operates powerfully towards changing their direction in seeking shelter on the coasts of Holland and France. 'During the presence of the herring and the mackerel in this latitude, their eggs may, during a calm, be seen floating on the surface of the water like sawdust, amidst an appearance like the wake or tract of a vessel, from which the course of the fish may be traced. Herrings generally disappear in this part of the channel about the beginning of December, and whilst along this coast are subject as well as the mackerel to a very formidable enemy in the dog-fish,* which are greatly increased within the last thirty years, a fact known to the cost of the fishermen, who have had their nets greatly injured by their quick cutting teeth.

Like the shark, the dog-fish turns on its side when it seizes its prey, resembling that ravenous fish in many respects; whenever it finds itself entangled in the net, it disengages itself in a few seconds by making a large incision, and passes through, probably liberating many herrings at the same time.

The dog-fish, in attacking the herring, devour them to repletion; they then disgorge what they have swallowed with great voracity, losing no time in recommencing seizing and swallowing the herrings with as much avidity as if it had been their first repast after a long abstinence, till they are again full, when their stomachs are again speedily relieved, which filling and emptying has continued with such perseverance as to exhaust the patience of the most curious observer. When this process is carried

* I may here observe that the herring gull (or *Laurus Fuscus*, of Linnæus), a bird about the size of the duck, is remarkable for its voraciousness, and particularly for devouring vast numbers of herrings. H. W. D.

on by numbers of the dog-fish about the nets, it occasions a white shining appearance on the surface of the sea, accompanied with smoothness, as if a quantity of oil had been strewed on it, emitting a rank oleaginous smell, which may be detected at some distance.

An idea may be formed of the numbers of the dog-fish about this channel, when it is stated, that, in the latter end of October, in the year 1827, some fishermen proceeded to a small sand-bank, situated about four miles to the east of Hastings, and two miles from land, in quest of cod-fish, and for this purpose shot lines, to which four thousand hooks were attached, over the ground. These, at the expiration of half an hour, were examined, when with very few exceptions a dog-fish was secured by every hook. A large cod had also been caught at the same time, but only the strong cartilages and bones of the head, with part of the vertebræ, remained, the rest having been swept away by the dog-fish : this was probably the work of only a few minutes after its capture. But their rapacity did not extend to their own species, the whole of which were hauled in uninjured. These insatiable fish are assisted in their ravages by the sepiæ, or cuttle-fish, which, with their hard mouths, resembling parrots' bills, cut up the mackerel and herrings with great adroitness. The sepiæ are sometimes attacked by the dog-fish, which they are generally enabled to frustrate by ejecting a liquid resembling ink, which renders the water turbid and obnoxious, and affords them an opportunity of making their escape.

The mackerel first met with off Hastings generally commence about the month of March, come from the German Ocean, to which they are supposed to belong, and appear to be of a different species from those caught off Mountsbay, in Cornwall ; the latter being longer, with the edges of the pectoral fins of a pink colour, and not so thick in proportion to the former, which are of a less weight, with the edges of the pectoral fins of a blue colour, and are considered of a superior quality.

The mackerel appear off Mountsbay always earlier than those off Hastings, and come from the Atlantic, remaining about a month or five weeks off Mountsbay, during which some decked fishing-boats from Folkstone, near Dover, proceed thither, and continue until the fish have disappeared. After an interval of a month, mackerel corresponding in every respect with those from the Atlantic appear off Hastings, by which it has been inferred that, after they have disappeared off Mountsbay, they take a south-easterly direction until they approach the coast of France, when they proceed to the east or north-east.

But the French fishing-boats, whose range of fishing-ground is very extensive, having never in the interval alluded to met with the Atlantic mackerel, which, before they make their appearance off this station, are invariably met with off Yarmouth and the North Foreland, this circumstance appears sufficiently conclusive that these fish proceed north about. The early mackerel are frequently accompanied by a few red multes (the salmonet of the Mediterranean); and whenever these nearly, if not altogether, equal the mackerel in number, the circumstance is generally the presage of the approach of great shoals of mackerel. The season for mackerel generally terminates at Hastings about the end of June or the beginning of July, although many have been caught in the middle and latter end of September, corresponding in appearance with those which appear off this place about the commencement of spring; and, as these are taken on the west side of the nets, it is concluded they are on their return to deep water in the German Ocean, leaving however some stragglers behind, which have been met with in the Channel the whole year.

ORDER.—ABDOMINALES.

GENUS.—SQUALUS.

SPECIES.

SQUALUS GRÖENLANDICUS,*

OR

GREENLAND SHARK.

UNTIL this animal (the Squalus Gröenlandicus) was noticed by Captain Scoresby, it had not been correctly described. This gentleman states it to be about twelve or fourteen feet in length, sometimes more, and from six to eight feet in circumference. The liver is very large, oily, and will fill a barrel. In its general form, it bears consi-

---

* SYNONYMES—The *Squalus Maximus*. LINNÆUS. *Squalus Borealis*. SCORESBY. *Squalus Gröenlandicus* vel *Arcticus*. DEWHURST.

The Squalus Gröenlandicus.—p. 246.

The abdominal surface of ditto.

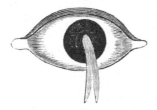

The appendage to the pupil of the eye.—p. 247.

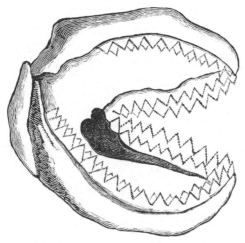

The Jaws of the Squalus Gröenlandicus.

derable resemblance to the dog-fish. The opening of the mouth, which extends nearly across the lower part of the under surface, is from twenty-one to twenty-four inches in width. The teeth are serrated in one jaw, and lancet-shaped and denticulated in the other. On each side, there are at least four or five rows, on one side so many as eight.

The eyes constitute the most extraordinary part of this animal. The pupil is of an emerald green colour; whilst the rest of the eye is blue. To the posterior edge of the pupil is attached a white vermiform substance, one or two inches in length. Each extremity of it consists of two filaments; but the central part is single. A representation of this singular appendage is given in the engraving.

The sailors imagine this shark to be blind, because it pays not the least attention to the presence of man; and is, indeed, so apparently foolish that it never draws itself back when a blow is aimed at it either with a knife or lance. We caught only one of these animals, which was on the 10th of May, 1824.

This shark annoys the whale exceedingly, and is one of its greatest foes. It bites it when alive, and feeds upon it when dead. And during the time the whalers are flensing a whale it is not uncommon to perceive a number of these animals in the neighbourhood, hence they frequently fall a sacrifice to their voracity.

It scoops large hemispherical pieces out of its body, nearly the size of a person's head; and continues scooping and gorging lump after lump, until the whole cavity of its belly is filled.

According to one observed by Crantz, he describes it of a cinerous grey colour, but which had a silverish appearance under water. The belly is of a much brighter and lighter hue. The skin is very rough, as if it was covered with coarse prickly grains of sand; and it is used for rasping and polishing wood, hence the sailors generally preserve the best of it. The head, which is two feet long and pointed anteriorly, although not very sharp, and the nostrils are placed below.

This fish has not the least perfect bone, but appears to belong to the *Chondropterygii,* or cartilaginous fishes, inasmuch as the whole skeleton (with the exception of the teeth, which are bone covered with enamel) is composed of that substance. In Norway and Iceland, the flesh is cut into long slices, and dried in the air for food. It generally brings forth four young ones at a time. When hoisted upon deck, it beats so violently with its tail, that it is dangerous to be near it, and the seamen generally dispatch it, without much loss of time. The pieces that are cut off exhibit a contraction of their muscular fibres for some time after life is extinct. It is, therefore, extremely difficult to kill, and unsafe to trust the hand within its mouth, even when the head is cut off. And, if we are to believe Crantz, this motion is to be observed three days after, if the part is trod on or struck.

When angling is employed for its capture, this author recommends an iron chain to be employed in lieu of a line, which it would either bite through or break. The Greenlanders strike it with a harpoon. It is said that this shark is very greedy of human flesh, and follows the ships in hopes of meeting with a corpse, and that it would sever the arm or leg of a seaman when in the water ;* but Captain Scoresby observes that although the whalers frequently slip into the water where sharks abound there has never been, to his knowledge, an instance wherein they have been attacked by this animal.† This animal is apparently so insensible to pain that although it has been run through the body with a knife, and has escaped, yet after a while it has been seen to return to banquet again upon the whale, at the very spot where it received its wounds.

---

* Crantz, Greenland, vol. i., p. 105.
† Scoresby's Arctic Regions, vol. i., p. 540.

The heart of this animal is very small, and performs from six to eight pulsations in the course of a minute, continuing its motions of contraction and dilatation for some hours, even when removed from the body.

Besides dead whales, the sharks feed upon small fishes and crabs. A fish, in form and size resembling a whiting, was found in the stomach of one killed by Mr. Scoresby; but the process of digestion had gone so far that its species could not be ascertained with any degree of satisfaction.

In swimming, the tail only is used, the rest of its fins being spread out to balance it, are never observed in motion but only when some change of direction is required.

The ventral fins of this animal are separate. It is without the anal fin, but has the temporal opening; it therefore belongs to the third division of this genus. The spiracles upon the neck are five in number on each side.

ORDER.—BRANCHIOSTEGI.

GENUS.—CYCLOPTERUS.

SPECIES.

CYCLOPTERUS LIPARIS.

OR

THE UNCTUOUS LUMP SUCKER.

WHEN perfectly fresh, the head and body of this fish are strongly marked with longitudinal streaks and waves of white, edged with blue, and disposed on a ground of testaceous or rather chestnut colour.

The body is naked, and the dorsal, anal, and caudal fin are united, forming its character. Mr. Donovan is of opinion that the Cyclopterus Lineatus of Iwan Lepechin is in

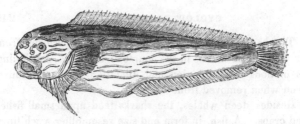

The Cyclopterus Liparis.—p. 249.

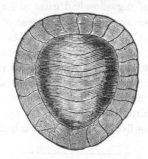

The Appendage of ditto.—p. 251

*The Cancer Ampulla.—*255.

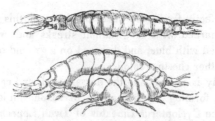

The Cancer Nugax—p. 256.

reality the same with the one here described. This naturalist failed in endeavouring to give his C. Lineatus such characteristics as would distinguish it from the liparis. And, on comparing Lepechin's description, we find them both, or very nearly alike. Mr. Donovan has observed the liparis to differ very considerably in its growth at various seasons of the year, and also in colour. Small specimens have occurred in which the sides and belly are white, in some a pale yellow, and in others rosy, the sides of the head usually partaking of the same tints as those of the body.

These fish resort in multitudes, during the spring, to the coast of Sutherland, near the *Ord* of *Cathness*. The seals, which swim beneath, prey greatly on them, leaving the skins; numbers of which, when thus emptied, float at that season ashore. It is easy to distinguish the place where seals are devouring this or any other unctuous fish, by a smoothness of the water immediately above the spot: this fact is now established, it being a tried property of oil, to still the agitation of the waves and render them smooth.* Great numbers of these fish, as also another species, the C. Lumpus or common lump-sucker, are found in the Greenland seas during the months of April and May, when they resort near to the shore for the purpose of spawning. Two of these were captured by Captain Phipps, during his voyage to the North Pole, to the northward of the Island of Spitzbergen.

The name of sea-snail is sometimes given to this fish, from the soft and unctuous texture of it resembling that of the land snail. It is almost transparent, and soon dissolves and melts away.† It is also found in the sea near the mouths of great rivers; Mr. Pennant states that he has seen it in January full of spawn.

* Philosophical Transactions, vol. lxiv. 1774, p. 445; and the London Mechanic's Register, vol. ii. 1826.
† Pennant's British Zoology, vol. iii. p. 180.   Donovan's British Fishes, vol. ii.

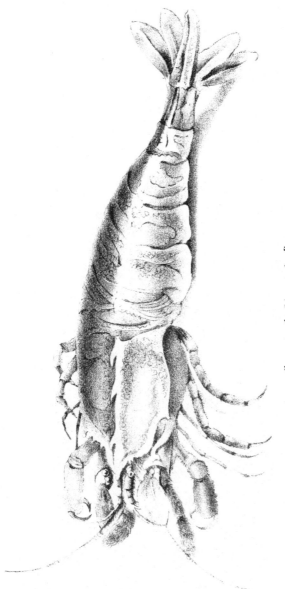

Cancer Arctica vel Boreas.

This animal, which is found on our coasts, seldom exceeds the length of four or eight inches, but such as frequent the shores of Greenland and Kamtschatka are oftentimes of a size far more considerable, being from a foot to eighteen inches in length.

The flesh of this fish is remarkably soft and oily; it is never eaten except by the inhabitants of Greenland, who devour it with avidity, who esteem it as highly nutritive and delicious.

The shape of the body of this fish is round, but near the tail it is compressed sideways; the belly is white and very protuberant; the head is large, thick, and round; there are no teeth in the mouth, but the jaws are very wide; the tongue is very large, the eyes small, the orifice to the gills also small; it has six branchiostegous rays; the pectoral fins are very broad, thin, and transparent, uniting almost under the throat; the first ray next the throat is very long, extending far beyond the rest, and is as fine as a hair; over the base of each is a sort of speculum or lid, ending in a point; this is capable of being raised or depressed at pleasure; behind the head begins the dorsal fin, which extends quite to the end of the tail; the ventral fin commences at the anus and unites with the other at the tail; beneath the throat is a round depression of a whitish colour, like the impression of a seal surrounded by twelve small pale yellow tubera, of which a representation will be found in the plate; in all probability it adheres to the stones by this apparatus like the other species.

In the dorsal fin there are thirty-six rays; in the pectoral thirty-two; in the anal twenty-six, and twelve in the tail.

In the winter 1803, Mr. Donovan accidentally detected a specimen of this fish, amongst a parcel of sprats brought for sale to the fish-market at Billingsgate.

CLASS.—CRUSTACEA.

THIS name, by which this class is distinguished, derives its
origin from *Crusta*, a *crust*, or *shell*, because the animals
have a covering of this kind.   The animals themselves are
familiarly known under the name of *crabs, lobsters, shrimps,
prawns, centipedes,* &c.   This were deemed by the ancients
as a sub-class of fishes, connecting the true fish with the
testaceous Vermes (*mollusca*); and this opinion prevailed,
with very little variation, as recent as the time of Linnæus;
who, in the great revolution which he effected in every part
of zoology, separated the *crustacea* from fishes and worms,
and placed them with insects.   After him, our industrious
countryman, Pennant, appears to be the first who separated
the crustacea from insects.   He has, however, neglected to
inform us his reason for this change, which renders it rather
an innovation than a reform, and deprives him of any claim
of priority which he otherwise might have deserved.   He
appears to have been rather influenced by caprice than by
any conviction of the correctness of his principles, and on
these grounds I shall not farther insist upon his claims.

The illustrious French zoologists, Baron Cuvier, Lamarck,
Latreille, and Dumeril, separated the *crustacea* from the
insecta, abandoning all the former opinions prevalent upon
the subject.   How far they may have been correct, in re-
jecting the doctrines sanctioned by many men of eminence,
remains to be ascertained.   Much caution is necessary in the
examination of innovations, and the utmost impartiality
should be used.   It is true that animals may have a decided resemblance in their external characters, whilst their
internal structure is totally different.   This has been the case
with the classes in question, although it appears to me as
very absurd to have placed together animals so very differ-

ent, and this was too frequently done by Linnæus. The most common observer would ridicule the idea that lobsters and crabs were insects! Yet Linnæus, and many specimen collectors of the present day, either from habit, or a veneration for Linnæus, still consider them as a branch of Entomology; and, as they both agree in having articulated limbs and antennæ, they are admitted by many British Entomologists into their cabinets as *genuine insects*, totally disregarding their internal structure, economy, and external appearance.

I shall now describe the distinctive characters of the Crustacea, as laid down by Cuvier, Dumeril, Latreille, and Lamarck. It appears that they agree with insects in having in common with them articulated limbs and antennæ. The *crustacea* respire by *gills* like the molluscæ, and have generally four *antennæ* or *horns*, and often six *mandibles* or *jaws;* likewise a *heart* similar to the molluscæ. They undergo little or no transformation, and, lastly, they breed more than once.

Such are the remarkable characters of this class, which appears to warrant a situation by itself. Indeed Linnæus himself, with that clearness and accuracy which distinguished his general views in every department of Natural History, has laid the foundation of those recent changes effected by the foreign zoologists. That great man has taught us to consider the internal organization " A natural, certain, unerring guide in the classification of animals." The changes thus effected will no doubt meet the views of all those who are competent to duly appreciate the true principles that should regulate every philosophical arrangement. The following are the characters of the Class Crustacea.

Anatomical Character.

Heart single; branchiæ or gills for respiration; no vertebræ; spinal marrow with many knots or ganglia; muscles for moving the feet.

External Characters.

A body with naked jointed feet, formed either for swimming or running; no wings; covering horny or crustaceous, horny, or membraneous, either shield-shaped or bivalve. Branchiæ or gills placed beneath the shell.

GENUS I.—CANCRI.

SPECIES I.

CANCER ARCTICA.*

THIS beautiful and singular species of the Genus Cancer was first discovered and described by Captain Phipps, who on making a dissection of a seal (*Phoca Gröenlandica*) found it in the stomach of that animal. The Captain placed it in the Systema Natura of Linnæus, after the Cancer Norwegicus.*

The thorax or chest is prickly; the second and third pair of legs are filiform; the proboscis is short, depressed, acute, and grooved on both the sides, having a very strong tooth underneath. This creature has two antennæ, and ten feet. Captain Scoresby informs us that he has occasionally found a similar species, to the one I have represented in the engraving. It is an inhabitant of the Northern and Greenland Ocean, but nothing further is known by naturalists respecting it; at least as far as I am aware of.

---

* Voyage towards the North Pole, p. 190.
† SYNONYMES—*Cancer Boreas.* PHIPPS. *Cancer Arctica.* DEWHURST.

SPECIES II.

CANCER PUSILLUS.

THE thorax or chest of this species is very round and entire; there is a tooth or projection on the tarsi or eyelids. It is greatly allied to the minutus, and is an inhabitant of the Northern Seas; it is extremely small, being not more than one-fourth the size of the minutus; it is also depressed and of a pale colour.

SPECIES III.

CANCER AMPULLA,

OR

THE BOTTLED-SHAPED CRAB.

THIS species possesses a head devoid of fangs, having fourteen legs. The late Earl of Mulgrave (then Captain Phipps) discovered one of these in the stomach of a seal, near the coast of Spitzbergen.* Captain Scoresby, Jun., found one in the stomach of a shark,† and, on opening the stomach of a female narwhale killed by our seamen in 1824, I likewise discovered one of these creatures.

---

* Voyage to the North Pole, 1774, p. 190.
† Scoresby's Arctic Regions, vol. i. p. 542.

SPECIES IV.

CANCER NUGAX.

Hands without fangs; legs fourteen in number; six posterior thighs, compressed and dilated. It chiefly inhabits the Northern Seas, and was captured in a trawl, by Captain Phipps, in 1774, near Moyen Island.

SPECIES V.

CANCER PULEX OF LINNÆUS.

The *Pulex Fluviatilis* of Ray. It has four hands, without fangs, according to Fabricius. It is an inhabitant of the Arctic Ocean, and like the preceding was taken in a trawl, near the coast of Spitzbergen. It is also commonly found in rivers, rivulets, and fountains: it swims in an incurvated posture, and is supposed to be luminous by night.

SPECIES VI.

CANCER MEDUSARUM.

This species has four hands, with a single fang; the head is very obtuse. According to Strom, it inhabits the Northern Seas, and is frequently found adhering to the medusæ: hence its appellation.

ORDER.—MALACOSTRACA.

GENUS II.—GAMMARINI.

SPECIES I.

CANCER vel GAMMARUS SQUILLA, OR SHRIMP.

This has a snout like a prawn, but deeper and thinner; the antennæ, or feelers, are longer in proportion to the bulk; the subcaudal fins are rather larger, and at full growth it is not more than one-half the size of the prawn. It is found near Spitzbergen, and about the Arctic shores.

SPECIES II.

CANCER vel GAMMARUS SERRATUS, OR PRAWN.

This species has a long serrated snout bending upwards; three pair of filiform feelers; claws small, furnished with two fangs; chest smooth; five joints to the tail; a fibulated middle caudal fin: the two outermost are flat and rounded. Captain Phipps found one of these in the stomach of a seal caught near Spitzbergen.

s

SPECIES III.

GAMMARUS ARCTICUS (LEACH),

OR THE

MOUNTEBANK SHRIMP OF SCORESBY.

THE characteristics of this animal are thus described by Dr. Leach. " G. Oculis sublunatis ; pedum pari tertio, secundo majori." It frequently tumbles over when in the water, with singular celerity, yet swimming with equal ease and rapidity in every position. The four feet are raised above the back, whenever that part of its body comes into contact with any solid substance. This species is discovered in all parts of the Spitzbergen sea, even at a great distance from land ; it also inhabits superficial water, and affords food for whales and birds.

SPECIES IV.

GAMMARUS MYSTICETUS,

OR THE

WHALE SHRIMP.

THIS is another species of this genus, to which I have given this name in consequence of having seen immense quantities of this species in the mouths and stomachs of the Balæ'na Mysticètus, during the period I was in Greenland. The only peculiarity otherwise worth notice is the largeness of its eyes.

SPECIES V.

ONISCUS CETI OF LINNÆUS,*

OR

WHALE LOUSE.†

THIS small animal, although only half an inch diameter, proves a source of great annoyance to the whole of the whale genus, by firmly attaching itself to the skin, and this principally under the fin, or any other part where the skin is tender and itself unlikely to be dislodged. Captain Scoresby states that he has only seen it in the B. Mysticetus; but from enquiries I have made to numerous Greenland Captains, and experienced seamen, I find that all the species are subject to them. The narwhale is liable to the annoyance of a similar but smaller animal.

CLASS —VERMES, OR WORMS.

LERNÆ BRANCHIATIS.

THIS is one of the largest species of this genus, being about two inches in length. The body is round and flexous, hollow and membraneous, thicker before and behind, it is of a dull white, or a dirty reddish hue. The mouth is lateral and seated between three slightly branched horns.

---

* SYNONYME.—The *Larundi Ceti*, of Dr. Leach.   † *Nom. Vulg.*

The neck is long, tubular, and filiform ; the tail ending in a perpendicular groove. It inhabits the Arctic Seas, and is found frequently adhering to the gills of the cod-fish, and to the lump-sucker (the Cyclopterus Liparis). Captain Phipps found one of them attached to the gills of the latter fish. The natives of Greenland and Davis's Strait frequently employ it as an article of food.

The Intestinal Worms, as the Ascarides, Echinorhyncus, Tænia, are frequently discovered in the alimentary canals of the different oceanic inhabitants of the Arctic Seas.

SIPUNCULUS LENDIX.

The body of this parasitical animal is cylindrical, with a subterminal aperture. This species, which I only find described by Captain Phipps, was found adhering to the intestines of an Eider duck. The late John Hunter dissected it, and informed Captain Phipps that he had seen the same species of animal adhering to the intestines of whales.

EXPLANATION OF THE ENGRAVING.

A. A portion of the intestine, with the animals adhering thereto.

B. One of the animals magnified.

C. The same cut open, showing the shape of the interior.

Captain Phipps, in a trawl on the northern side of Spitzbergen, found two species of the Genus Ascidiæ, viz. the *Gelatinosa* and *Rustica.* They belong to the Mollusca tribe, having a body which is fixed, roundish, and apparently issuing from a sheath ; the apertures are two, generally placed near to the summit, one below the other. These creatures are more or less gelatinous, and have the power of contracting and dilating themselves at pleasure.

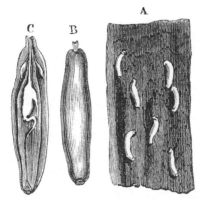

The Sipunculus Lendix.—*p.* 260.

The Synoicum Turgens.—*p.* 284.

CLASS.—MOLLUSCÆ.

INTRODUCTORY OBSERVATIONS.

ALTHOUGH, as respects the external configuration of this class, the general plan of their organization is not as uniform as that of the vertebrata, yet there is always an equal degree of resemblance between them in their structure and functions. Their nervous system consists of two long cords, running longitudinally through the abdomen, dilated into knots, or ganglia; the first of these knots is placed over the œsophagus or gullet, and is denominated " brain: " it is scarcely larger than those are along the abdomen, with which they communicate by filaments that encircle the œsophagus like a necklace. The covering or envelope of the body is divided by transverse folds into a certain number of rings, whose teguments are sometimes soft and sometimes hard; the muscles, however, being always situated internally. Articulated limbs are frequently attached to the trunk; but very often there are none. These animals were named by the late Baron Cuvier " *Animalia Articulata*," or articulated animals, in which is observed the transition from the circulation in closed vessels to nutrition by imbibition, and the corresponding one of respiration in circumscribed organs, to that effected by trachead or air-vessels distributed throughout the body. In them the organs of taste are the most distinct, one single family alone presenting that of hearing. Their jaws, when they have any, are always lateral.*

* CUVIER, *Regne Animal*, vol. i. p. 24.

TESTACEOUS MOLLUSCÆ.

GENUS.—CLIO.

SPECIES I.

CLIO, HELICINA,

OR

THE SEA PEARL. *

THIS constitutes a species of the Order Mollusca, according to Gmelin, who denominated
it the *"male slime fish;"* it is covered with a
beautiful delicate shell, not dissimilar in form
to the *nautilus*. In diameter it measures from
two-eighths to three-eighths of an inch. It is
found in immense quantities near the Spitzbergen coast, but
is rarely discovered out of sight of land. Captain Phipps
captured two of them in a trawl, near Seven Island Bay.

SPECIES II.

CLIO-RETUSA.

THE annexed representation of this animal
is taken from the late Mr. O'Reilly's work
on Greenland, and belongs to the same order
as the preceding, but appears to be destitute

---

* SYNONYMES.—*Argonautica* of Fabricus. *Clio Helicina*, Linnæus
and Phipps.

of all testaceous envelope; according to this author it forms
one of the principal mollusca, which constitutes the food
of the Balæ'na Mysticètus.

SPECIES III.

CLIO ARCTICA.*

THESE creatures are found in immense numbers about Spitzbergen, but are not universal throughout the Polar Seas.
In swimming, it brings the tips of its fins almost in contact
with each other, first on the one side and then on the contrary. Captain Scoresby, Jun., kept several of them in a
tumbler of sea water for about a month, when they gradually
wasted away and died. Martin says they form part of the
food of the B. mysticètus, and hence they are denominated
" *Whale's food.*"

ORDER.—CEPHALOPODA.

SPECIES.

SEPIA GRÖENLANDICA,

OR

THE GREENLAND CUTTLE FISH.

As this species of the cuttle fish appears not to be so well
known as the other varieties, and as no distinguishing characteristic name has been given to it, I have therefore ventured
to designate it as above, from the portion of the Arctic

---

* SYNONYMES.—*Clio Limacina*, Linnæus. *Clio Borealis*, Scoresby.
Sea May Fly, Marten. *Clio Arctica*, Dewhurst.

Ocean in which it is discovered. I have unfortunately not been successful in procuring a representation of it, but I may observe that it differs but little from the *Sepia Octopadia*. It is discovered in great numbers in many parts of the Greenland Seas; and, from having procured many from the stomachs of the narwhale, I am inclined to believe that it forms a part of the food of that species of cetàcea.

ORDER I.—MULTIVALVES.*

GENUS.—CHITON.

SPECIES.

CHITON RUBER,

OR

COAT OF MAIL SHELL.

THE shell of this animal is oblong, elevated on the back, of a reddish hue; it is variegated with eight valves, divided on each side, from the anterior margin to the beak, into two compartments, the anterior transversely situated, the striæ of which bend and cross the posterior compartment; the spaces between the striæ are broad; the border is rough; it barely measures more than half an inch in length, or more than a quarter in breadth.† Captain Phipps took some of these in a trawl on the north side of Spitzbergen, and is also common in Scotland on rocks at low water.‡

---

* Section III. Gen. LIII. *Brewster's Conchology.*
The classification and numerical arrangement of the Testaceous Mollusca here adopted is the same as that followed by Sir David Brewster, in the excellent article " *Conchology,*" inserted in his Cyclopedia.
† Ibid.
‡ *Lin. Syst. Nat.* 1107, 7. *Brewster's Cyclopedia,* vol. vi. Part i. p. 102.

ORDER II.—BIVALVES.

DIVISION I.—DENTATED.

GENUS.—MYA, OR THE OYSTER.*

SPECIES FOUND IN GREENLAND, HAVING GAPING VALVES.

MYA, OR PEARL OYSTER.

GENERIC CHARACTERS.—Shell generally gaping at one end, and furnished with broad, strong, thick, broad teeth at the hinge, and not inserted into the opposite valve.

SPECIES I.

MYA TRUNCATA,

OR

TRUNCATED PEARL OYSTER.†

THE shell is truncated at the smaller end, where 'it gapes, and is wrinkled concentrically; the valves are concave and reflected at the smaller end, both broad and erect. It is about two inches long, and about three broad.‡ Captain Phipps found this species on the beech in Smeerenberg Harbour. It is also found lodged under gravel at low water mark, on various parts of the British shores. It is the shell which Pennant should have referred to as containing the animal eaten by the Hebridians. In Orkney and Shetland it is used as a supper dish when boiled, and is called " smurslin."

---

* Genus xlvi. *Brewster's Conchology.*

† SYNONYME.—*Avicula Margaritifera.*

‡ *Pennant's British Zoology,* vol. iii.

As the pearl oyster is an inhabitant of the Polar as well as of other seas, it may not be deemed uninteresting to the scientific as well as the general reader to be made acquainted with the manner in which the pearl is secreted by this animal; for the following lucid description, which I here insert, I am indebted to a valuable contemporary.*

OBSERVATIONS ON THE FORMATION OF PEARL.

" Pearl is a calcareous secretion of molluscous animals deserving notice. It is secreted only by the fish of bivalves, and principally by such as inhabit shells of foliated structure, as sea and fresh water muscles, oysters, the Pinnæ, &c. A pearl consists of carbonate of lime, in the form of nacre, and animal matter arranged in concentric layers around a nucleus. Each layer is presumed, but I know not on what grounds, to be animal; so that a pearl must be of slow growth, and those of large size can only be found in full-grown oysters. ' It is the nacral lining of the central cell that produces the lustre peculiar to the pearl, which cannot be given to artificial ones.'

" Pearls, as Mr. Gray justly observes, are merely the internal nacred coat of the shell, which has been forced, by some extraneous cause, to assume a spherical form. They are, therefore, not properly ' a distemper in the creature that produces them,' and cannot, under any view, be compared with Calculi in the kidney of man;† for, though accidental formations, and, of course, not always to be found in the shellfish which are known usually to contain them, still they are the products of a regular secretion, applied, however, in an unusual way, either to avert harm or allay irritation. That in many instances they are formed by the oyster, to protect itself against aggression, is evident; for, with a plug of this nacred and solid material it shuts out worms and other intruders which have perforated the softer

---

* *Magazine of Natural History*, vol. v. for 1832.
† Lister, *Hist. An. Ang.* p. 150.

shell, and are intent on making prey of the hapless inmate : and it was apparently the knowledge of this fact that suggested to Linnæus his method of producing pearls at pleasure, by puncturing the shell with a pointed wire. But this explanation, it is obvious, accounts only for the origin of such pearls as are attached to the shell; while we know that the best and the greatest number, and, indeed, the only ones which can be strung, have no such attachment, and are formed in the body of the animal itself. ' The small and middling pearls,' says Sir Alexander Johnston, ' are formed in the thickest part of the flesh of the oyster, near the union of the two shells ; the large pearls almost loose in that part called the beard.' * Now these may be the effect merely of an excess in the supply of calcareous matter, of which the oyster wishes to get rid ; or they may be formed by an effusion of pearl, to cover some irritating and extraneous body. The reality of the latter theory is, perhaps, proved by a practice of the modern Chinese, who force the swan muscle (*Anodon cygneus*) to make pearls, by throwing into its shell, when open, five or six minute mother-of-pearl beads strung on a thread : in the course of one year, these are found covered with a crust which perfectly resembles the real pearl. The extraneous body which naturally serves for the nucleus appears to be very often, or, as Sir E. Home says, always, a blighted ovum. Christophorus Sandius, in 1673, on the authority of Henricus Arnoldi, ' an ingenious and veracious person,' asserted that the ova left unexpelled from the shell became the nuclei on which pearls, in the fresh water muscle, were formed. ' Sometimes,' he says, ' it happens that one or two of these eggs stick fast to the sides of the matrix, and are not voided with the rest. These are fed by the oyster against her will ; and they do grow, according to the length of time, into pearls of sufficient bigness, and imprint a mark both on the fish and the shell, by the situation, conforming to its figure.' This theory has been

† *Sir Everard Home's Lectures on Comparative Anatomy*, vol. v. p. 308.

fully adopted by Sir E. Home, from whose paper I have made the above quotation. ' If,' says the enthusiastic Baronet, ' I shall prove that this, the richest jewel in a monarch's crown, which cannot be imitated by any art of man, either in the beauty of its form or the brilliancy and lustre produced by a central illuminated cell, is the abortive egg of an oyster enveloped in its own nacre, of which it receives annually a layer of increase during the life of the animal, who will not be struck with wonder and astonishment ? * And, as proofs of this, he informs us that he has always found the seed-pearls in the ovarium, or connected with that part of the shell on which the ovarium lay; and he has discovered that all Oriental pearls have a brilliant cell in the centre, of a size exactly large enough to contain one of the ova. ' From these facts, I have been led to conclude that a pearl is formed upon the external surface of an ovum ; which, having been blighted, does not pass with the others into the oviduct, but remains attached to its pedicle in the ovarium, and, in the following season, receives a coat of nacre at the same time that the internal surface of the shell receives its annual supply. ' This conclusion,' he adds, ' is verified by some pearls being spherical ; others having a pyramidal form, from the pedicle having received a coat of nacre as well as the ovum.'†

" I will conclude what I have to say concerning pearls with the following extract from the paper of Mr. Gray, quoted in the preceding page :—' The pearls are usually of the colour of the part of the shell to which they are attached. I have observed them white, rose-coloured, purple, and black ; and they are said to be sometimes of a green colour. They have also been found of two colours; that is, white with a dark nucleus, which is occasioned by their being first formed on the dark margin of the shell before it is covered with the white and pearly coat of the disk, which, when it

* Sir Everard Home, vol. v. p. 302.

† *Philosophical Transactions*, 1816.   Part iii. p. 339.

becomes extended over them and the margin, gives them
that appearance.

" ' Pearls vary greatly in their transparency. The pink
are the most transparent; and in this particular they agree
with the internal coat of the shell from which they are
formed; for these pearls are only formed on the Pinnæ,
which internally are pink and semi-transparent, and the
black and purple specimens are generally more or less
opaque.

" ' Their lustre, which is derived from the reflection of
the light from their peculiar surface produced by the curious
disposition of their fibres, and from their semi-transparency
and form, greatly depends on the uniformity of their texture and the colour of the concentric coats of which they
are formed. That their lustre does depend on their radiat-
ing fibres may be distinctly proved by the inequality of
the lustre of the ' Colombian pearls,' which are filed out of
the thick part near the hinge of the pearl oyster (Avicula
margaritifera), so that they are formed, like that shell, of
transverse laminæ, and they consequently exhibit a plate of
lustre on one side which is usually flat, and are surrounded
by brilliant concentric zones, which show the places of the
other plates, instead of the even, beautiful, soft lustre of
the true pearls.' "

SPECIES II.

MYA ARCTICA,

OR

THE ARCTIC OYSTER.

The shell of this animal is striated, having the valves with
two subspinous ridges, and the hinge without teeth. This
oyster is discovered in the Arctic shores, generally among
a species of Algæ ; it is about the size of a bean, of a palish
yellow colour externally, and of a milk-white interiorly.*

* _Encyclopedia Londinensis_, vol. xvi. p. 428.

ORDER III. EQUIVALVES.

DIVISION II. EDENTATE, OR TOOTHLESS.

GENUS.—MYTILUS, OR THE MUSCLE.*

SPECIES DISCOVERED IN THE ARCTIC SEAS.

SPECIES I.

MYTILUS RUGOSUS vel ARCTICUS,

OR

THE POLAR MUSCLE.

THE shell of this species is rhombic, brittle, rugged, and
rounded at the ends. It inhabits the northern seas and
lakes. It was found by Captain Phipps in Smeerenberg
Harbour. Great numbers of them usually are found lodged
in limestone, each in a separate apartment, with apertures
which are too small for the shell to pass through without
breaking the stone. The shell is about an inch and a
quarter long, and in breadth about half its length; its
colour is of a dirty grey, and within is half blue and half
white, marked with very small longitudinal striæ, crossing
the transverse wrinkles.

SPECIES II.

MYTILUS FABA, OR YELLOW MUSCLE.

THE shell of this species is oval, yellow, striate, with a
cremulate margin. It inhabits the Greenland seas, attaching itself to the rocks, by a bronzed byssus or beard.

---

* Genus xlvi. *Brewster's Conchology.*

GENUS.—MEDUSA.*

GELATINOUS ANIMALS OF THE ARCTIC REGIONS.

I AM sure there are but few of my readers who have visited the ocean, but must have had opportunities of seeing specimens of this very extraordinary genus of animals, which have the appearance and consistence of a whitish gelatinous substance, of a round form, and when left by the sands dissolve into a fluid.

It was to this remarkable genus of animals that Linnæus applied the name of Medusa. They are more numerous than they have hitherto been supposed, and, by the peculiarities of structure, this family is increased to more than a hundred and fifty species. The following is a description of their paradoxical singularities.

The generic characters of these animals are a gelatinous, orbicular body, flat generally underneath; mouth central beneath. They consist mostly of a tender gelatinous transparent mass, of different figures, furnished with arms or tentacular processes, proceeding from the lower surface. The large species when touched cause a slight tingling and redness: hence they are called sea-nettles. They shine with great splendour in the water. The form of the body differs in different species, but when at rest is generally the segment of a sphere; they swim well, and appear to perform that motion by rendering their body more or less convex, and thus striking the water. When left on shore they are motionless and appear more like flat gelatinous cakes than living animals.

---

* SYNONYME.—Sea Blubber, or Jelly Fish. Sea Nettles. *Nom. Vulg.*

Mons. Peron, during his voyage to the South Seas, collected a great number of this remarkable genus of animals.
In a memoir presented by him to the National Institute of
France, in 1809, their singularities are thus expressed.
" Their substance seems to be merely a coagulated water ;
yet the most important functions of life are exercised in it.
Their multiplication is prodigious ; yet we know nothing of
the peculiar mode in which it is effected.   They are capable
of attaining the weight of fifty or sixty pounds, and of becoming several feet in diameter; yet their nutritive system
escapes our eyes.   They execute the most rapid and long-
continued movements; yet the details of their muscular
system are imperceptible.   They have a very active species
of respiration, the true seat of which is a mystery.   They
appear extremely feeble ; yet fish of considerable size form
their daily prey, and dissolve in a few moments in their
stomach.   Many species of them shine in the night like balls
of fire, and some sting or benumb the hand that touches
them; yet the principles and agents of both these properties remain to be discovered.   If put into fresh water, they
die in about half an hour." *

All the medusæ have a gelatinous body, nearly resembling
the cap of a mushroom, which the Abbé Spallanzani denominated " *Umbella;*" but they differ, some in wanting or
some in having a mouth ; in the mouth being simple or
multiplicious; in the presence or absence of a production
resembling a pedicle ; and in the edges of this pedicle, or
of the mouth itself, being furnished with tentacula, or filaments, more or less numerous.   Some of these animals exhibit beautiful colours.   They are found in all climates, but
more particularly in the Southern Ocean.

Some of the species have the power of benumbing the
hand when touched; hence the name of *sea nettles* have
been applied to them.   The appearance of many is pe-

---

* *Zool. Dan.* ii. t. 77.

culiarly graceful and elegant when floating in their native
element, from the delicate colours with which they are
adorned. The bodies of some among them are of a light
azure blue, the border surrounded with the appearance of
golden beads like a coronet, from which stream in every
direction delicate threads of a bright carmine colour; in
short, almost all those that are found in warmer climates have
something pleasing to the spectator, either in form or colour.

It is to the scientific observations of Captain Scoresby,
Jun., that zoologists possess any information of these inhabitants of the Frozen Seas; for previous to his interesting
investigations in 1816, on the sea water, they were not supposed to exist.

The economy of these little creatures, as constituting the
foundation of the subsistence of the largest animals in the
creation, has already been mentioned. The common whale
feeds upon them, and perhaps these again on the minor
medusæ and animalculæ. The finned whales and dolphins
devour principally herrings and other small fishes. The
bear's general food is probably the seal, and the seal subsists
on the cancri and small fishes; these again prey on others
smaller to themselves. Thus we find the whole of the
larger animals dependent on these minute beings, which
establish a chain of existence; and were not the whole
chain beautifully supported, and one single link to give
way, the stupendous fabric must inevitably perish.

It is the medusæ and animalculæ that give the green-olive
hue to the Arctic Ocean, and the number has been computed
to be almost beyond the power of human calculation.*

As Captain Scoresby is the only author who has given
any account of these animals, I must extract a few observations respecting them from his "Account of the Arctic
Regions." The names which head them, I have given, as
this gentleman had not appended any to his accurate descriptions.

---

* Vide, page 24.

T

SPECIES I.

MEDUSA OVIFORMA,

OR

EGG-SHAPED MEDUSA.

REPRESENTED the natural size. An extremely sensitive animal; when touched, or indeed when the vessel in which it is contained is moved, it shrinks into an irregular globular mass. It is divided into eight segments, by as many rows of finny fringes. These, though only perceptible by their iridescence when in motion, are capable of moving the animal through the water. Its colour is greyish-white, but reddish (pale lake red) in the longitudinal cavity. Is found in the Spitzbergen Sea.

SPECIES II.

MEDUSA CUCUMIS,

OR

APPLE-SHAPED MEDUSÆ.

OBLONG, with eight ciliate ribs, iridescent fringes, with a single cavity. It is found in Spitzbergen and Greenland Seas; moves very slowly by means of the fibres on the ribs, and, when touched, it contracts into the form of an apple. The body is white mixed with blue, and covered with irregular red spots.

SPECIES III.

MEDUSA OVOIDALIS,

OR THE

OVAL-SHAPED MEDUSA.

FORM ovoidal. Eight segments. A double cavity united
by a small canal. Sensitive Iridescent fringes. Found
in latitude 75° 40′ N. Longitude 5° or 6° W. Colour
similar to the former.

SPECIES IV.

MEDUSA PILEUS.

THIS is one of the most curious medusæ. It consists of eight
lobes, with a beautifully iridescent finny fringe on the external edge of each. A canal, four-fifths of the animal,
penetrates the centre of it; and two red cirrhi, which may
be extended to the length of nearly a foot, proceed from
a crooked cavity in opposite sides. The animal is semi-
transparent. Its colour is white with a blush of red, the
finny fringes of deeper red. It is found of various sizes;
one specimen taken up in latitude 75° 40′ N., longitude
5° or 6° W., in a green-coloured sea, was three inches in
length. It also inhabits the Mediterranean Sea, and is perhaps the luminous species described by Forskal, under the
name of *Medusa Densa*.

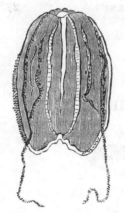

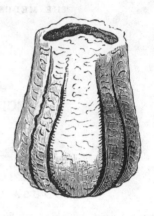

The Medusa Pileus.—p. 276.   *The Medusa Marsupialis.*—p. 276.

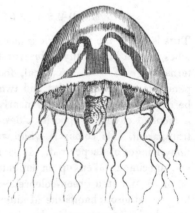

The Medusa Utriculus.
p. 276.

The Medusa Campanula.
p. 277.

SPECIES V.

MEDUSA MARSUPIALIS,

OR

PURSE-SHAPED MEDUSA.

THE substance of this is tougher than any other species previously examined. It has one large cavity, and is divided by the finny fringes into eight segments, each alternate pair of which are similar. The colour is of a very pale crimson, with waved purple lines; finny fringes deeper crimson. This animal appeared almost without sensation. The only evidence it gave of feeling was in an increased vibration of the finny fringes. Though it was cut into pieces, each portion, on which there was any fringe, continued, by its incessant play, to give evidence of life during two or three days; after which it became putrescent and began to waste away. This animal was found in the Spitzbergen Sea. Captain Scoresby never saw but one specimen.

SPECIES VI.

MEDUSA UTRICULUS,

OR

BOTTLE-SHAPED MEDUSA.

CAUGHT in a transparent green sea, in latitude 75° 48' N., longitude 8° W. A sensitive animal. Large single cavity. Eight segments. Finny fringes white and iridescent. Form ovoidal, with compressed mouth.

SPECIES VII.

MEDUSA AURANTII SCORESBY*I*,

OR THE

ORANGE-COLOURED MEDUSA,

OF SCORESBY

THIS singulra species Captain Scoresby had only the opportunity of seeing one, and after whom I have named it. It was sent him by the late Captain Bennet of the Venerable of Hull, being found in latitude 75° 20′ N., longitude 11° 50′ E. On the right extremity (vide cut) there was a transverse slit or opening. This animal was convex above and concave beneath. The length was three inches, breadth nearly an inch; its thickness one-third of an inch. When slit open, it exhibited a number of transverse bands and three cavities. Its colour was a brilliant orange. It was not transparent, nor tenacious of life, having died to all appearance soon after it was taken.

SPECIES VIII.

MEDUSA CAMPANULA,

OR

BELL-SHAPED MEDUSA.

THE disk of this Medusa is gibbous, the border white and ciliate; beneath which is a hairy cross. The body is conicorbicular, beneath hollow and snowy; the fringe of the margin and cross yellow : the latter is often white ; it is an inhabitant of the Greenland Seas.

CLASS.—ANIMALIA RADIATA,

OR

RADIATED ANIMALS.

The Creator's glorious works around us
Are his proofs of kindness, and of love to man :
Alike evidences of his power and wisdom,
Each silently persuades us to adore
That goodness of a God, who hath formed
Them for us, by whom they were
Destined for our use, and admiration
Of a Being, powerful, wonderful and great.

THIS forms a class of Radiated Animals. In the three first
great divisions of the Animal Kingdom, by the late Baron
Cuvier,* the organs of sense and motion are symmetrically
arranged on the two sides of an axis. There is a posterior
and anterior dissimilar face. In this last division, they are
disposed like rays round a centre, and this is the case even
where they consist of but two series; for then the two faces
are similar. They approximate to the homogeneity of
plants, having no very distinct nervous system or particular
organs of sense ; in some of them it is even difficult to discover a vestige of circulation ; their respiratory organs are
universally seated upon the surface of the body, the intestine in the greater number is a mere sac without issue, and
the lowest of the series are nothing but a sort of homogeneous
pulp, endowed with motion and sensibility.†

---

* Viz. 1. Animalia Vertebrata. 2. Animalia Mollusca. 3. Animalia
Articulata. 4. Animalia Radiata.

† Vide the translation of the *Règne Animal*, vol. i. p. 24. Now publishing by Henderson, of the Old Bailey, London.

GENUS.—ASTERIÆ, OR STAR FISHES.*

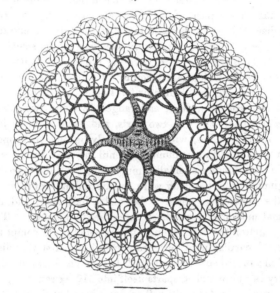

SPECIES I.

ASTERIAS CAPUT MEDUSA,

OR THE

MAGELLENIC STAR FISH.†

THE body of this curious creature is divided into five parts or rays, as may be seen in the engraving, which represents

---

* SYNONYME.—Sea stars.

† SYNONYMES.—*Asterias Caput Medusæ* Habitat in mari Norwegieo. LINN. SYST. NAT. *Asterias Radiata*, radiis duplicata dichotomis. FABR. FR. GRÖENLANDICA. It is called the Medusa's Head, because it bears some resemblance to the snaky hair, which painters have made peculiar to Medusa.

the belly and under part of the animal. In the centre are
observed some hollow rings, which it is supposed constitutes the mouth, since they are furnished with pores through
which the creature receives its nourishment in the water.
To the under part of the five rays are affixed a great
number of legs, so disposed that by their help the creature
contrives to walk, although very slowly. In the space between is a black thick skin, covering the back or upper part
of the body, which is divided by ribs, from the centre of
which grow out five double or ten single arms or rays, of a
yellow or reddish colour, and ranged two and two. The
rays are subdivided into two parts, and each of these into
two other branches, which successively become in proportion smaller and more numerous, until the ray terminates in
an infinite number of small ramifications. All the branches
beginning from the body were, in the animal represented
in the engraving, about a foot and a half long, and the
animal measured upwards of three feet in diameter. The
body and all the rays, even to the extremity, are composed
of hard cartilaginous vertebræ, resembling a star or disk,
and are furnished with a prickle. It is to be remarked that
every ray ultimately disparts itself into 512 extremities ; the
five rays therefore altogether are separated into 2560.
Now, every branch, having 512 extremities, is composed
of 1023 joints ; the five rays therefore consist of 5115. As
every ray contains upon an average six vertebræ, there are
altogether no less than 81,840 cartilaginous vertebræ ; when
therefore one of these creatures comes to its full and perfect
growth, the joints, extremities, and vertebræ, produce a
number almost incapable of being counted. As the ancients
believed this creature to be a vegetable, doubtless they supposed that it absorbed its nourishment by pores, similar to
the flowers that grow in water.   But Linnæus and all
modern zoologists find its mouth very distinctly in the centre,
where five valves are united in a point. All the claws and
joints are movable, and lie extended in the water, thus

making the animal appear like a flower in full bloom. It is frequently found upon a rock, or twisted upon a coral. In order to effect its capture, they touch them with a stick, upon which they immediately adhere. When drawn out of the water, they hang down loosely like a bundle of flax, and, when put on a table to be examined, they contract themselves and become hard. The fragments of their rays furnish the fossil entrachi. If we drown this creature in spirits of wine, and keep the rays flat and expanded in the execution, it is easy to extract, by means of a pair of forceps, the stomach of the animal whole and entire through the mouth. Respecting the difference between the large and small species, the one has the claws less divided, and the other them all forked. This species is found in the Indian Ocean, at the Cape of Good Hope, in the Arctic Seas, and near Archangel; it is said to be found of a prodigious size. The Caspian Sea, however, abounds with those of the largest dimensions. This creature appears to form the medium by which nature passes from naked worms to testaceous animals, and shell-fish in general.*

Sir Arthur de Capell Brooke procured two which were captured on the Norwegian coasts, and, as he saw no others, he considered them as somewhat rare. He observes,† that the smallest of them so exactly resembles the engraving of that described by Pontoppidan, under the name of the arborescent or star-fish, that it would induce a man to form a favourable idea of the general accuracy of the figures in his work. The largest, when alive and expanded, must have been of a considerable size, was drawn up by the nets of the fishermen ; and the merchant, Mr. Buck, in whose possession it was, and who had hung it up in his house as a curiosity, was kind enough to let him have it. With regard to its habits, food, and other particulars relating to this singular animal, hardly any thing is known ; nor are we assisted in

---

* Wilke's London Encyclopedia, vol. ii. p. 801.
† Travels in Sweden, Norway, and Finmark, p. 326.

this respect by Linnæus, Fabricius, or other authors. It certainly is by no means common on the coast of Norway and Lapland, though it has been generally supposed to be found in abundance there. Pontoppidan, speaking of the *krake* or *kraken*, says: " it seems to be of that polypus kind which is called by the Dutch *ze sonne*, by Rondeletius and Gesner, *stella arborescens*, i. e. a star which shoots its rays into branches like those of trees; according to the same exact description, I gave it the name of *Medusa's Head.* A very worthy person told me he had seen some of them of an extraordinary bigness, and others have seen them four times as large as the common size, splashing the water about with their numerous branches or arms."*

The gallant author of an interesting article in the United Service Journal for July 1833, p. 331, respecting Captain Ross's voyage to the Arctic Seas in 1818, states that one of this species of *Asteriæ* was captured in consequence of its clinging to a " *deep sea-clam*,"† that was sent down for the purpose of sounding in 1000 fathoms (6000 feet) water, near the south side entrance of Lancaster Sound.

SPECIES II.

ASTERIAS RUBENS COMMUNIS,

OR THE

COMMON STAR FISH.

THE whole substance of this animal appears like an assemblage of bones in the form of wedges, which in their size and figure resemble small brushes, and are supposed to be the young of the parent animal, every one of which as they increase in bulk, and approach to maturity, fall off from the

---

* Bishop Pontoppidan's Natural History of Norway.

† This ingenious instrument was invented by Captain Ross, for the purpose of ascertaining the nature of the earth, sand, small shells, &c., lying at the bottom of the vast abyss, and is a species of claw-forceps.

Asterias Rubens.

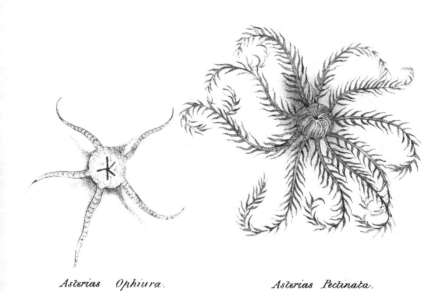

Asterias Ophiura.   *Asterias Pectinata.*

Friedel. lith 24 Greek St Soho Sqʳ

parent, and become a separate asteria. If any of the rays
or arms are broken off, they regenerate, and grow again,
which fact may be seen in the engraving. The creature
had by some accident lost the fifth ray; but the accretion
of a new one, small, and just growing out, is plainly to be
seen in the plate. Found by Captain Phipps at Spitzbergen.

SPECIES III.

ASTERIAS OPHIARA,
THE LIZARD TAIL OR WORM STAR-FISH.

THIS species is very common in the North Seas; but the
one represented was found in the American Ocean. The
upper surface of this animal is composed of vertebral rings,
communicating a power of motion to the rays on every side.
They are membranaceous and very tender. Its colour is
blue in the natural state, but when dried becomes ash-
coloured. This species is found on rocks, and between
lumps of coral, to which it firmly adheres. It has very
quick motion.

SPECIES IV.

ASTERIAS ARANCIA.

THIS is the *Astropleten Echinatus Minor* of Mr. Link. The
structure of this individual differs materially from the others,
although the rays spring from a point in the centre, and the
bones are placed in rows like the preceding, but more numerous, and differing from each other in size. This species is
not rough, and the bones are laid flat by the sides of each
other, being only separated by deep notches. In the centre
part, where the five rays unite, the last vertebra of two of

them, at the points where they are opposed to each other, form an oblong aperture to each ray; there are therefore five of them. The upper surface of this species also differs from the other varieties: and nothing but a similar vertebra to the one just mentioned is to be observed; only that those of the under surface turn a little round the side and then become jagged. Respecting the other exterior points they are all beset with an infinite number of hairy pilli thickset and erected, like the surface of coarse plush. Every one of these erected points is supposed to be a new animal growing out of the old.

One of this species of Asterias came up with the lines which were hauled aboard when we lost the whale of the seventh of July 1824, in lat. 79° 37' N.* It was of a brownish red colour, and about three inches and a half in diameter.

SPECIES V.

ASTERIAS PECTINATA.

THIS was likewise taken by Captain Phipps in a trawl, on the northern side of the coast of Spitzbergen.

CLASS.—ZOOPHYTA.

SPECIES.

SYNOICUM TURGENS.—PHIPPS.

THIS species of animal flower was captured in a trawl on the north side of Spitzbergen, by Captain Phipps, who considered it as new to zoologists, and totally different from the Zoophytes hitherto known and described, so that he deemed it a distinct genus. He was of opinion that naturalists should place it next to the Alcyonium, with which it in some particulars agree, but differ from it materially in having the openings for the animals only at the top, and the ani-

* Vide, page 60.

mals themselves not exserted like the polypes (*Hydra*), which is the case in the Alcyonium.*

CLASS.—ANIMALCULÆ,

OR THE

ANIMALCULÆ OF THE POLAR REGIONS.

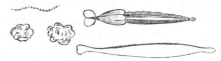

THESE constitute a class of very diminutive animals, which in general are invisible without the aid of a microscope. They are, however, usually divided into three distinct sections, viz. the *visible, microscopical,* and *invisible :* the first, though visible, cannot be accurately discerned without the help of glasses ; the second are only discoverable by the microscope ; and the last are merely presumed to exist, as they are still unknown. The existence of the latter cannot well be disputed, though it cannot be asserted, unless we conclude that the microscope has arrived to the highest degree of perfection. Reason and analogy give some support to the conjectures of naturalists in this respect; animalcules are discerned of various sizes, from those which are visible to the naked eye to such as appear only like moving points under the most powerful miscroscopic lenses ; therefore, it is not unreasonable to imagine, therefore, that there are many others that still resist the action of the microscope, as the fixed do that of the telescope with the greatest powers hitherto invented.

Among the visible animalculæ are included an amazing variety of creatures by no means of an analogous nature. Those numerous animals which crowd the water in the summer months, changing it sometimes to a deep or pale red

---

* Captain Phipp's Voyage, p. 199.

colour, green, yellow, &c., are of this description. The
larger kinds are chiefly of the insect or vermes tribes, and
of which one, denominated the *monoculus pulex*, is sometimes so abundant as to change the water apparently to a
deep red. A similar appearance is likewise caused by the
cercaria mutablis, when it váries from a green to a red
colour ; the *vorticella fasciculata* also changes it to a green,
and the *rotatoria* to yellow. To this section also must be referred many of the acaca and hydrachna genera.

The microscope discovers numberless myriads of these
diminutive creatures to our view in most fluids, among which
I may mention sea-water, the colour of which I have reason
to believe is given by the animalculæ; fresh river or rain
water, the animal fluids, vinegar, beer, dew, &c. Also in
animal or vegetable infusions, and many of the chalybeate
waters.

Those who have made the most minute researches, and
accurate enquiries into the natures of the several objects subjected to their senses, have found that the substances upon
which they employed their curiosity were often quite different to what at first they appeared to be. Thus, for instance, the whole earth has been replenished with an inexhaustible store of the least of which we should the least
suspect; that is, an infinite number of animalculæ floating
in the air we breathe, sporting in the fluids we drink, or
adhering to the several objects which we see and handle.
The conjectures and theories relating to the production,
generation, structure, and uses of these animals, have been
as various as were ever contrived by caprice, or embraced
by credulity. Not to bewilder the reader, however, in these
labyrinths, but to prove the truth of this assertion of these
animals' existence, we have recourse to the microscope, by
which we are not only able to distinctly see them, but in
some degree to distinguish their shapes and peculiarities of
motion, particularly by Lieutenant Drummond's and Holland's Oxy-Hydrogen Microscopes, Carpenter's Solar

Microscope (in Regent Street, Piccadilly), and several others of a similar nature exhibiting in London. Mr. John Varley, Mr. Baüer (of Kew), and Mr. George Francis (of Berwick Street, London), have constructed powerful instruments of this nature.

The contemplation of animalculæ has made the idea of infinitely small bodies extremely familiar to us. A mite was by the ancients considered as the smallest animal in existence; but naturalists are not now surprised to be told of animals *twenty-seven millions of times smaller than a mite.*

Minute animals are proportionably much stronger, more active and vivacious than large ones. The spring of a flea in its leap, how vastly does it outstrip any thing the greater animals are capable of! A mite, how vastly faster does it run than a race-horse! Mons. de L'Isle has given the computation of the velocity of a little creature scarcely visible by its smallness, which he found to run three inches in half a second. Now, supposing its feet to be the fifteenth part of a line, it must make five hundred steps in the space of three inches; that is, it must shift its legs five hundred times in a second, or in the ordinary pulsations of an artery.* A similar fact is mentioned by that delightful author and my kind friend, Mr. Sharon Turner, in his invaluable work;† one which has afforded me the greatest pleasure in its perusal.

The excessive minuteness of microscopical animalculæ conceals them from the human eye. One of the wonders of modern philosophy is the invention of the means to produce instruments, whereby the perception of these creatures may be brought under the cognizance of our senses: an object of a thousand times too little to be able to affect us. Yet we have extended our views over animals to whom these would be mountains. In reality, most of our microscopical animalculæ are of so small a magnitude, that through

* Hist. Acad. 1711, p. 23.

† The Sacred History of the World, as displayed in the Creation, and subsequent Svents to the Deluge, attempted to be philosophically considered in a Series of Letters to a Son, by Sharon Turner, F. R. S. Fourth Edition. Longman and Co.

a lens whose focal distance is the tenth part of an inch, they appear as only so many points; that is, their parts cannot be distinguished, so that they appear from the vertex of that lens under an angle not exceeding an object If we investigate the magnitude of such an object, it will be found nearly equal to $\frac{3}{100000}$ of an inch long. Supposing, therefore, these animalcules of a cubic figure, that is, of the same length, breadth, and thickness, their magnitude would be expressed by the cube of the fraction $\frac{3}{100000}$, that is, by the number $\frac{27}{1000\,000\,000\,000\,000}$, that is, so many parts of a cubic inch is each animalcule equal to.*

Leeuwenhoeck calculates that a thousand millions of animalculæ in common water are not altogether so large as a grain of sand.

SPECIES.

The animalculæ of the Arctic Seas have not been much investigated by zoologists; but the following, which are mentioned by Captain Scoresby, will give the reader some idea of these creatures :—

Fig. 1, when examined in the field of a double microscope, appeared of the size of a grain of coarse sand.   It was of a brownish colour, and its movements were in a direct line.

Fig. 2 was about half the dimensions of the preceding; its configuration approached that of a globe; it was of a dark colour, with a species of tail, and in its movements it advanced in a curious zigzag direction.

Fig. 3 was considerably smaller; it moved about with amazing rapidity, by sudden starts, pausing for an instant between each locomotive impulse, and then springing into a new direction.

---

* Rees's Cyclopedia, vol. ii.

Having concluded these remarks, I may, in terminating my zoological description of the oceanic inhabitants of the Arctic Regions, not inappropriately quote the eloquent language of Professor Lyall* on this subject; he observes that " the ocean teems with life—the class of polypi alone are conjectured by Lamarck to be as strong individuals as insects.   Every tropical reef is described as bristling with corals, budding with sponges, and swarming with *crustacea, echini,* and *testacea,* whilst almost every tide-washed rock is carpeted with *fuci,* and studded with *coralines, actiniæ,* and *molluscæ.*   There are innumerable forms in the seas of the warmer zones, which have scarcely begun to attract the attention of the naturalist ; and there are parasitic animals without number, three or four of which are sometimes appropriated to one genus, as to the Balæ'na, for example. Even though we concede, therefore, that the geographical range of marine species is more extensive in general than that of the terrestrial (the temperature of the sea being more uniform, and the land impeding less the migrations of the oceanic than the ocean of the terrestrial), yet we think it most probable that the aquatic species far exceed in number the inhabitants of the land.   Without insisting on this point, we may safely assume, as we before stated, that, exclusive of microscopic beings, there are between *one and two million of species* now inhabiting the terraqueous globe ; so that if only one of these were to become extinct annually, and one new were to be every year called into being, more than a million of years would be required to bring about a complete revolution in organic life."

CONCLUSIONARY REFLECTIONS.

I HAVE now brought to a conclusion my description of the inhabitants of the Northern Seas; and whether we consider them in a philosophic or commercial point of view, yet to the reflective mind I believe I have pourtrayed sufficient

* Of King's College, London.

U

in the preceding pages for the reader to look up towards
an Omnipotent Creator,

" In whom we live and have our being,"

with feelings of humility and pure devotion ; for it will be
seen I have attempted to delineate some of the greatest and
some of the most diminutive objects of his creative power :
and as all the living beings inhabiting this globe are made
more or less subservient to the wants of man in all situations,
climates, and latitudes, therefore well might the poet
Thompson justly exclaim,

" 'Tis surely God,
Whose unremitting energy pervades,
Adjusts, sustains, and agitates the whole ;
He ceaseless works alone, and yet alone
Seems not to work, yet with such perfection fram'd
Is this complex stupendous scheme of things."

The contemplation of animated nature generally is (taking
it as a science) replete with so many points of interest, it
embraces so many topics by the investigation of which useful knowledge is acquired, our intellectual character improved, and a degree of real pleasure not to be calculated
is felt and enjoyed, so that although zoologists generally
make choice of one department of Natural History for their
minute study and investigation, yet it is difficult which to
select, as the best calculated to satisfy the curious, to gratify
the philosopher, or to repay for the outlay of time and research the mere utilitarian.

Dr. Robertson,* of Chesterfield, justly observes that,
" whatever be the department which may seem most calculated to effect so much and to interest so many, however
high the pre-eminence which it may have attained, either
with reference to the interesting or the useful nature of the
facts which it developes, still such subject is capable of
further improvement. Without altering in any important

---

* Lectures on the mutual relations between the vital functions in
animals and plants, p. 7.

degree the body of the picture, its colouring may be improved: we may likewise with reference to any department
of science, by conjoining with it another but analogous investigation, add to such research the lovely and beauteous
shadings of variety and contrast."

It is by thus connecting and coupling different departments of science, as anatomy and zoology for example,
and thus uniting branches of natural philosophy which are
too frequently considered separately, and as if they had no
connection with each other, that we are enabled to view as
it were by a single gigantic effort of mental vision the
whole of the animal kingdom, as one beauteous and har-
monious picture; it is thus we are enabled to see all
natural objects, whether animal, vegetable, or mineral,
mutually dependent on each other; it is thus we view
nature as a whole and not as a part, each portion of which
must have been created with an evident design; and, if we
admit the existence of this, we must likewise admit that of
the designer, who from the overwhelming magnitude of the
work of creation must be Omnipotent, who from the
harmony and mutual dependence of the whole must be
Omniscient, and from the carefulness with which every
thing occurs throughout the universe must be Omnipresent,
and from the union of these this wondrous designer can be
none other than God, the Lord of Hosts. On this subject
the poet may not unaptly be quoted, who thus observes :—

> " Still may I note how all the agreeing parts
> Of this consummate system join to frame
> One fair, one finished, one harmonious whole ;
> Trace the close links which form the perfect chain
> In beautiful connection ; mark the scale
> Whose nice gradations with progression true,
> For ever rising, end in Deity."

But, to go still further, when we reflect on the wondrous
and beauteous harmony, the awful sublimity, the majestic
grandeur which is displayed in every part of the creation,
how much must onr ideas of the power of that Being, of the

faculties of that mind, which formed all these be increased! How heartfelt and unassumed a sense of humility, a virtue which may be justly considered the most distinguished ornament of man, must take the place of those presumptuous ideas and opinions which are too often to be found in the minds of those who are unfortunately ignorant of the great truths that are only to be found in the book of nature, and which it is the business of science to reveal.

The meanest insect we can see, nay, the minutest and most contemptible weed we can tread upon, is really sufficient to confound Atheism,* and baffle all its pretensions. How much more astonishing is that variety and multiplicity of God's works with which we are continually surrounded! Let any man survey the face of the earth, or lift up his eyes to the firmament; let him consider the nature and instincts of the lower animals, and afterwards examine the operations of his own mind; will he then presume to say or to suppose that all the objects he meets with are nothing more than the result of unaccountable accidents and blind chance? Can he possibly imagine that such wonderful order should spring out of confusion? Or that such perfect beauty should even be formed by the fortuitous operations of unconcious, inactive particles of matter? As well, nay, better, and more easily, might we suppose that the earthquake might happen to build towns and cities; or the materials carried down by a flood, fit themselves up without human hands into a regular fleet. For what are towns, cities, or fleets, in comparison with the vast and amazing fabric of the Universe? The answer is promptly given: *Nothing.*

The pious Bishop Watson remarked, he had long thought the motions of the heavenly bodies, the propagation and

---

* I did not conceive that there was such a being as an Atheist existing; but unfortunately, at the last November Old Bailey Sessions, two misguided individuals avowed themselves confirmed *Atheists:* earnestly do I hope they may be convinced of their unbelief, and returned to the paths of religion and happiness.

growth of animals and plants, the faculties of the human
mind, and even the ability of moving our hands in any one
direction, by a simple volition, afford, when deliberately
reflected on, to constitute more convincing arguments
against Atheism than all the recondite lucubrations of the
most profound philosophers; and in this opinion do I most
cordially coincide. For I may remark, in a word, the argu-
ment for the existence of a God, which is drawn from a
contemplation of nature, is so clear and so strong that the
most ignorant can comprehend it, and the most learned
cannot invent a better.

> " To study God, God's student MAN, was made
> To read him as in nature's text convey'd,
> Not as in heav'n; but as he did descend
> To earth, his easier book: Where to suspend
> And save his miracles, each little flower,
> And lesser fly, shows his familiar power."

The cultivation of the science of Natural History not only
refines the man, but dignifies and exalts the affections. It
elevates them to the admiration and love of that Being who
is the author of every thing that is fair, sublime, and good in
the universe. Scepticism and irreligion are rarely com-
patible with the sensibility of heart which arises from a just
and lively relish of the wisdom, harmony, and order sub-
sisting in the world around us; and emotions of piety must
spontaneously arise in every bosom that is in unison with all
animated nature. Actuated by this divine inspiration, man
finds a fane in every grove; and, glowing with devout fer-
vour, he joins his song to the universal chorus, or muses the
praise of the Almighty in more expressive silence. Thus
they

> " Whom nature's works can charm, with God himself
> Hold converse; grow familiar day by day,
> With his conceptions; act upon his plan,
> And form to his the relish of their souls."

And on reviewing the works of the creation which are more
or less daily presented to our view, we can come to this

only conclusion; first, that they all equally proclaim the existence of a God, for

> " Each shell, each crawling insect, holds a rank
> Important in the plan of Him who form'd
> This scale of beings, holds a rank which lost,
> Would break the chain, and leave a gap
> Which nature's self would rue."

And, secondly, that every branch of Natural History is fully capable of yielding us innumerable objects for the formation of a pleasing, scientific, and even religious study. Its chief tendency ought to lead us from the admiration of the works, to the serious contemplation of their author. In fact, to teach us to look through nature up to nature's God. It is a study terminating in the conviction, the knowledge, and adoration of that all-gracious and merciful Being, to whose bounteous beneficence alone we are all indebted for every blessing we enjoy.*

---

* To such of my readers as are interested in the study of the works of of the Creator, I may recommend to their attentive perusal, the volume I have already mentioned by my kind and learned friend Mr. Sharon Turner, and likewise " The Christian Philosopher," and other works, by my talented friend the Rev. Dr. Dick of Dundee. These valuable productions prove their authors to be not only profound philosophers, but also Christians, in the most extended meaning of the word.

THE END.

APPENDIX.

I HERE subjoin, for the perusal of my Subscribers, a selection from the numerous testimonials I have been honoured with from time to time, by some of the most talented men in the three kingdoms; and I trust that my humble claims to the patronage of the literary and scientific public are not made without some solid foundation.

TESTIMONIALS, CERTIFICATES, LETTERS,

&c. &c.

Blenheim Street, Anatomico-Chirurgical Society.

LABORE ET HONORE.

TO Mr. HENRY WILLIAM DEWHURST,

THIS testimonial of esteem for his professional abilities, and approbation of the zeal he has upon all occasions evinced for the welfare of this Society, is unanimously voted at a Meeting held this 14th day of December, 1822.

Signed by order of the Society,

G. W. HUME, President.
RICHARD MACKRELL, Secretary.

LONDON VACCINE INSTITUTION,

FOUNDED UNDER THE MAYORALTY OF

SIR JAMES SHAW, Baronet, M. P., Vice President.

OPIFERI QUE PER ORBEM TERRARUM IMUS.

On the Recommendation of the BOARD OF MANAGERS, founded on the Report of the MEDICAL COUNCIL, the LONDON VACCINE INSTITUTION receives H. W. DEWHURST, Esq., Surgeon, &c., into the number of its HONORARY MEMBERS.

JOHN WALKER, M.D.
Director.

The GOVERNORS respectfully request MR. DEWHURST's *acceptance* of this, their DIPLOMA, in testimony of the high value they place on *his liberal co-operation* with them, in the philanthropic cause of VACCINATION.

HUGH BEAMS,
Secretary.

Domine Dirige Nos.

LONDON, 14th MARCH, 1823.

The Right Honourable the LORD MAYOR,

WILLIAM HEYGATE, M. P., President.

St. GEORGE'S HOSPITAL,*

FOR the sick and lame. Supported by voluntary contributions and benefactions of the nobility, gentry, and others.

This is to certify that Mr. HENRY WILLIAM DEWHURST hath *very diligently* attended the PRACTICE OF SURGERY at St. GEORGE'S HOSPITAL for *Twelve Months*, and has been for two months an attentive dresser.

Witness our hands, this 19th day of July, 1823.

Signed. EVERARD HOME,
JOHN GUNNING,
ROBERT KEATE,
B. C. BRODIE,
} Surgeons.

---

* I am happy to state that this Hospital has been rebuilt, and I trust the patrons of this work, will not consider this valuable Institution unworthy their support and patronage.

𝕿𝖍𝖊𝖘𝖊 𝖆𝖗𝖊 𝖙𝖔 𝖈𝖊𝖗𝖙𝖎𝖋𝖞 that Mr. Henry William Dewhurst hath *regularly and diligently* attended *Four* Courses of my Lectures on Anatomy, Physiology, and Surgery, and hath carefully dissected the Human Body.

Dated this 24th day of December, 1823.

JOSHUA BROOKES, F. R. S. & F. L. S.,
Soc. Cæs. Nat. Cur. Mosq. Soc. &c. &c.

Theatre of Anatomy, Blenheim Street,
Great Marlborough Street.

The writer of the following note is the mustering officer appointed by His Majesty's Government, who examines the Testimonials of all the Surgeons engaged on board vessels in the Merchant service.

This note procured me an appointment as Surgeon of the Neptune, of London, Matthew Ainslie, Commander, to whom it was directed.

Custom-House, London.

Sir,
I have seen Mr. Dewhurst's papers, which are satisfactory.

M. P. ROUBILLIARD.
Dec. 31, 1823.

𝕿𝖍𝖎𝖘 𝖎𝖘 𝖙𝖔 𝖈𝖊𝖗𝖙𝖎𝖋𝖞, that Mr. H. W. Dewhurst has been receiving my instructions in the Practice of Physic, Materia Medica, Pharmaceutical Chemistry, &c. &c., for the space of six months.

JOHN HARDING, Surgeon,
Member of the Apothecaries' Hall, and Apothecary to the St. George's and St. James's General Dispensary,
No. 60, King Street, Golden Square.

London, Jan. 30, 1824.

London, February, 1824.

𝕿𝖍𝖊𝖘𝖊 𝖆𝖗𝖊 𝖙𝖔 𝖈𝖊𝖗𝖙𝖎𝖋𝖞, that Mr. H. W. DEWHURST hath *diligently* attended our Lectures on the Theory and Practice of Midwifery, and on the Diseases of Women and Children, during *Five Courses.*

RICHARD BLAGDEN.
T. A. STONE.

ROYAL INFIRMARY FOR DISEASES OF THE EYE,

MARY-LE-BONE STREET, PICCADILLY.*

𝕿𝖍𝖎𝖘 𝖎𝖘 𝖙𝖔 𝖈𝖊𝖗𝖙𝖎𝖋𝖞, that Mr. H. W. DEWHURST has very diligently attended me for nearly two years, during which period he has acquired the ART OF CUPPING, and has himself under my inspection *cupped generally the different parts of the Human Body*, (the TEMPLES more particularly), and I do consider him sufficiently qualified to practice the said art of Cupping, with dexterity and neatness.

Witness my hand, this 2d day of March, 1824.

C. LEESE,
Cupper to the above Infirmary, &c.
H. W. Dewhurst, Esq.

ROYAL INSTITUTION OF GREAT BRITAIN,

ALBEMARLE STREET.

London, February, 1824.
Mr. DEWHURST has attended a course of my Lectures on Chemistry.
(Seal.)   WILLIAM THOMAS BRANDE.

St. GEORGE'S AND St. JAMES'S DISPENSARY,

No. 14, OLD BURLINGTON STREET†

MISERIS SUCCERRERE DISCO.

𝕿𝖍𝖎𝖘 𝖎𝖘 𝖙𝖔 𝖈𝖊𝖗𝖙𝖎𝖋𝖞, that Mr. HENRY WILLIAM DEWHURST hath attended *with great diligence* one course of my Lectures on the MATERIA MEDICA, delivered at this Dispensary.

GEORGE GREGORY, M.D.,
Physician to the Small Pox Hospital, &c. &c.
London, the 26th day of March, 1824.

----

* This excellent Institution is removed to King William Street, Charing Cross, and is denominated the "Royal Westminster Ophthalmic Hospital," the talented Surgeon of which is G. J. Guthrie, Esq., F. R. S., President of, and Professor of Anatomy to the Royal College of Surgeons.

† The Dispensary has removed from Old Burlington Street, to King Street, Golden Square.

St. GEORGE'S AND St. JAMES'S DISPENSARY,

No. 14, Old Burlington Street.

MISERIS SUCCERRERE DISCO.

𝕿𝖍𝖎𝖘 𝖎𝖘 𝖙𝖔 𝖈𝖊𝖗𝖙𝖎𝖋𝖞, that Mr. HENRY WILLIAM DEWHURST hath attended *with great diligence*, two courses of my Lectures on the Theory and Practice of Physic, delivered at this Dispensary.

GEORGE GREGORY, M. D.,
Licentiate of the Royal College of Physicians of London, and Senior Physician to the Dispensary.

London, the 26th day of March, 1824.

St. GEORGE'S AND St. JAMES'S DISPENSARY,

No 60, King Street, Golden Square.

MISERIS SUCCERRERE DISCO.

𝕿𝖍𝖎𝖘 𝖎𝖘 𝖙𝖔 𝖈𝖊𝖗𝖙𝖎𝖋𝖞, that Mr. HENRY WILLIAM DEWHURST, hath diligently attended the Practice of the Physicians, for a period of nine months at this Dispensary.

GEORGE GREGORY, M. D., PHYSICIAN,
Licentiate of the Royal College of Physicians of London.

London, the 26th day of March, 1824.

𝕿𝖍𝖊𝖘𝖊 𝖆𝖗𝖊 𝖙𝖔 𝖈𝖊𝖗𝖙𝖎𝖋𝖞, that Mr. HENRY WILLIAM DEWHURST served me as an apprentice, agreeably to the articles, expressed in his *Indentures* from the 1st day of August, 1818, to the 1st day of August, 1824, during the whole of which time he gave me entire satisfaction by his assiduity, attention, and general good conduct.

Given under my hand, this 2d day of August, 1824.

THOMAS WALLINGTON, SURGEON, R. N.

BAYSWATER.

WHEN the Council of the University of London issued their advertisements for Professors to fill the Chairs for that Establishment, in reply to a letter addressed to the late Mr. Brookes, I had the honour to receive the following answer.

Blenheim Street, Great Marlborough Street,
29th January, 1828.

MY DEAR SIR.

I very much lament that your application came too late for me to benefit you by a recommendation, having already sent into the Council of the London University a testimonial for another Candidate,* in whose favour an application was solicited a week ago.

I remain, my dear Sir,

With best wishes, yours ever,

JOSHUA BROOKES.

ROYAL JENNERIAN SOCIETY,

MDCCCIII.

"Kings shall be thy nursing Fathers, and their Queens thy nursing Mothers." Isaiah xlix. 23.

Names of the Patrons.†

Honorary Diploma.

London, March, 1828.

Under the Presidency of Field-Marshal His Grace the DUKE OF WELLINGTON, *Cuidad Rodrigo*, and *Vittoria, Prince of Waterloo*, K.G., G.C.B. &c. &c.

At the General Court of the Royal Jennerian Society.

The Director, DOCTOR WALKER, *in the Chair.*

The Governors, by an unanimous Vote, did themselves the high gratification to

Elect H. W. Dewhurst, Esq.,

An Honorary Member of their Great Royal Establishment.

JOHN FOX, SECRETARY,

22, Bridge Street, Blackfriars.

ANDREW JOHNSTONE,
 REGISTRAR AND SUB-TREASURER,
 52, Burr Street.

J. W. CHAIRMAN.

Old Jewry.

This is to certify, that I have known Mr. HENRY WILLIAM DEWHURST, *Surgeon* and *Teacher* of *Anatomy* and *Surgery*, for the last four years, and from my intimate acquaintance with him, I believe him to be a young man possessing an extensive knowledge of his profession, and that he bears an unexceptionable moral character.

Witness my hand, this 8th day of June, 1828.

JAMES ROSE, NOTARY PUBLIC.

---

* I may observe, that Mr. Brookes never gave but one recommendation for any vacant office, and this accounts for his refusal.

† I have omitted the names of the Patrons of this Society, as most of them kindly patronized this work.

68, *Hatton Garden, Oct.* 1, 1828.

My Dear Sir,

I have to apologize for not replying to your note at an earlier period, and beg to assure you that nothing could give me more sincere pleasure than advancing the interests of any member of the profession.

I feel complimented by your request, that I would visit your Museum, and certify my opinion of it, which I shall do as soon as convenience shall permit. From the great industry and deep research you have shown in your Dictionary of Anatomy and Physiology, *the only one in any language,* and in your other works, I am disposed to think, that the legal authorities of the Royal College of Surgeons and Apothecaries' Hall, will readily agree as to your competency to discharge the duties of Lecturer on Anatomy and Surgery.

I remain, dear Sir, yours very truly,

M. RYAN, M. D. M.R.C.S. L. & E.

Lecturer on Medicine, Midwifery, and Medical Jurisprudence, also Editor of the London Medical and Surgical Journal.

Sir,

I feel much pleasure in forwarding to you the enclosed Copy of a Resolution agreed to by the Committee of the Southwark Literary and Scientific Institution.

I am, Sir, your obedient servant,

S. HARRISON, Hon. Sec.

Secretary's Office, Union Street, Borough,
15th November, 1828.

At a Meeting of the Committee of the Southwark Literary and Scientific Institution, held on Friday, November 14th, 1828,

It was Resolved unanimously,

That Professor Dewhurst be elected an Honorary Member of this Institution, as an acknowledgment for his able services as a Lecturer on Popular Anatomy, and for his zealous assistance in promoting the interests and success of this Institution.

S. HARRISON, Hon. Secretary.

This is to certify, that H. W. DEWHURST, Esq., Lecturer on Human and Comparative Anatomy, Zoology, &c. &c., is perfectly qualified to teach, and that he possesses every means of doing so : a Museum fast approaching to perfection, casts, plates, &c., and it is not in my power to speak too highly of his abilities and qualifications. I therefore most strongly and earnestly recommend him as a teacher.

Signed.   H. W. BULL,
                                        Member of the Royal College of Surgeons
                                        in London, and Surgeon Royal Navy.
Binfield near Bracknell, Berks.
    Nov. 25th, 1828.

___

𝕴 𝖉𝖔 𝖍𝖊𝖗𝖊𝖇𝖞 𝖈𝖊𝖗𝖙𝖎𝖋𝖞, that I have known Mr. HENRY WILLIAM DEWHURST for upwards of *seven years*, and have had numerous opportunities of witnessing his talents, both as a Surgeon and a Lecturer on Anatomy, Physiology, and Surgery, and from my knowledge of these important requisites in a medical education, 𝕴 𝖉𝖔 𝖈𝖊𝖗𝖙𝖎𝖋𝖞, that he possesses both means and abilities for teaching these sciences. He also possesses a good collection of Preparations illustrative of Health and Disease, casts, drawings, and engravings, by the most eminent continental anatomists, which, with a recent subject, are sufficient for the completion of a course of Lectures on these important subjects.

    Witness my hand this 13th day of December, 1828.

                                       C. LEESE.

18, Princes Street, Cavendish Square.
To H. W. DEWHURST, Esq

___

                    York Place, Edinburgh,
                            August 18, 1829.

𝕴 𝖇𝖊𝖌 𝖑𝖊𝖆𝖛𝖊 𝖙𝖔 𝖈𝖊𝖗𝖙𝖎𝖋𝖞, that HENRY WILLIAM DEWHURST, Esq., Surgeon, M. W. S., Lecturer on Human and Comparative Anatomy, and author of several scientific and valuable Publications in the Medical branches of Education, is a gentleman perfectly qualified to instruct in the department which he professes, in testimony whereof, I beg leave further to observe, that I have myself reaped from attendance on his Lectures many valuable and instructive hints, from which I have derived much sterling and practical knowledge.

           HENRY F. W. VON HEYDELOFF, M. D.,

        Member Extraordinary and late President of the Council of the
        Royal Physical Society, Edinburgh; also Member of the
        Hunterian Medical Society, and late one of the Assistant Phy-
        sicians to the Edinburgh Lying-In Institution.

HAVING had in contemplation about four years ago, in conjunction with Mr. Cherry, of Clapham, to establish a new Veterinary School, I wrote to Professor Coleman, in order that that eminent gentleman might not deem me an opponent* in this department of Science, when I was honoured with the following letter.

ROYAL VETERINARY COLLEGE, *27th August,* 1829.

MY DEAR SIR,

I have the honour to acknowledge the receipt of your letter, and I beg to assure you, that I have no objection whatever to the formation of one or more Veterinary Schools in London. In talents, in knowledge, in zeal; and industry, I am aware that you and your friends will be formidable opponents but I have no desire to monopolize all the Veterinary students and I shall never oppose you, or any other gentleman,† who without depreciating the characters of o ers, seeks honourably to establish his own reputation by his own merits.

I have the honour to be, Sir,

Your most obedient humble servant,

EDWARD COLEMAN.‡

To H. W. DEWHURST, Esq.

FROM SIR JAMES M'GRIGOR.

ARMY MEDICAL DEPARTMENT, *December,* 1829.

DEAR SIR,

I have been duly favoured with your letter of the 25th ult., and am aware of your competency as a teacher on the subject, but medical officers of the army are not required to engage in those studies.§

I have the honour to be, dear Sir,

Your most obedient humble servant,

J. M'GRIGOR, DIRECTOR GENERAL.

BEING a Candidate for the Office of Secretary to the Royal Humane Society,‖ in 1830, the two followin testimonials were given to me, in addition to the others I then produced.

---

* I may observe that my intended colleagues were bitterly opposed to that gentleman from private motives, while I had every reason to be otherwise ; this will explain the reason of my writing to him on the subject

† Since this was written, my friend Professor Youatt has ably succeeded in establishing a Veterinary class in the University of London.

‡ This gentleman is the Professor of Veterinary Anatomy, Physiology, and Pathology, to the College. He is the author of a splendid work on the Structure and Economy of the Horse's Foot ; and I have on several occasions experienced the liberal kindness of this truly talented individual, which I have much pleasure in publicly acknowledging.

§ Veterinary and Comparative Anatomy, Zoology, &c. &c.

‖ From what I have since heard, I should have succeeded in obtaining

FROM JOHN WILSON, ESQ., A. M.,
Lecturer on Natural Philosophy.

January 1, 1830.

𝕿𝖍𝖎𝖘 𝖎𝖘 𝖙𝖔 𝖈𝖊𝖗𝖙𝖎𝖋𝖞, that from my long and intimate acquaintance with Mr. HENRY WILLIAM DEWHURST, I consider him to be a person qualified in every respect for the office of *Secretary to the Royal Humane Society.*

Witness my hand,

JOHN WILSON, A. M.

Vernon House Academy,
32, *Queen's Row, Walworth.*

————

14, *Bryanstone Street, Portman Square.*

𝕴 𝖍𝖊𝖗𝖊𝖇𝖞 𝖈𝖊𝖗𝖙𝖎𝖋𝖞, that I have known Mr. H. W. DEWHURST for several years, and from my own personal knowledge, believe him to be a young man possessing an excellent knowledge of general business, and of his profession as a Surgeon, also as the author of many valuable publications connected with medical education, and possessing a good moral character. I therefore consider him perfectly qualified to hold the office for which he is a Candidate, as he is remarkable for his perseverance and assiduity.

Witness my hand, this 1st day of January, 1830.

J. M. ROSE, A. M., M. R. C. S. E.,
SURGEON AND LECTURER ON CHEMISTRY.

————

From the same.

London, Jan .1, 1830.

𝕿𝖍𝖎𝖘 𝖎𝖘 𝖙𝖔 𝖈𝖊𝖗𝖙𝖎𝖋𝖞, that HENRY WILLIAM DEWHURST, Esq., attended in the year 1824, two courses of my Lectures on Experimental Chemistry, with great diligence.

J. M. ROSE, A. M.,M. R. C. S. E.,
SURGEON, LECTURER ON CHEMISTRY..

————

London, January 5, 1830·
*Great Portland Street.**

IN reply to Mr. DEWHURST's request of Sunday last, I hereby attest his having been my Dispenser, &c., for about four years, during which time he conducted himself with honesty, civility, and sobriety.

WILLIAM HOPEFULL LEREW.

———————————————————————————

this office, had not the wealthy relatives of the fortunate candidate anticipated me, by making themselves governors, and of course they voted in favour of their friend.

* Now of No. 60, Norton Street, Portland Place.

FROM GREVILLE JONES, Esq., Surgeon,
Professor of Anatomy and Surgery, and late Editor of the Medical
Examiner.

8, Hatton Garden.

I have much pleasure in stating, that from the conversations I have had with Mr. Dewhurst, and from perusing his works, I consider him a highly intelligent practitioner.

Signed.   GREVILLE JONES.
January 23, 1830.

This is to certify, that Mr. H. W. DEWHURST hath *regularly and diligently* attended eight courses of my Lectures on Anatomy, Physiology, and Surgery, *and three courses of Dissections. Mr. D. hath also carefully dissected the Human Body thrice.*

Dated this 26th day of July, 1830,

JOSHUA BROOKES, F. R. S., F. L. S., F. Z. S.*

Soc. Cæs. Nat. Cur. Mosq. Soc., &c. &c.

Theatre of Anatomy, Blenheim Street,
    Great Marlborough Street.

Dublin, March 16, 1831.

31, York Street.

Mr. Houston † presents his compliments to Dr. Dewhurst, and feels much obliged by his polite attention in sending him a copy of his *ingenious " Nomenclature for the Sutures of the Cranium,"* and begs to assure him that the principles of his improvements are such as to merit his entire approbation.

To Dr. Dewhurst, London.

Dublin, March 12, 1831.

Harcourt Street.

Mr. Kirby‡ presents his compliments to Dr. Dewhurst, and feels honoured by his kind attention in sending him his Synoptical Table of im-

---

* Mr. Brooks's second Certificate for an attendence on eight additional courses of Lectures to those mentioned at page 297.

† This gentleman is Demonstrator of Anatomy to the Royal College of Surgeons in Dublin.

Now Dr. Kirby, professor of Anatomy and Surgery to the Royal College of Surgeons in Dublin.

x

proved Nomenclature for the Sutures of the Cranium. The best testimony Mr. Kirby can bear to its value, is that he will adopt it in his class. It is easily pronounced and clearly expresses all it is intended to convey. Mr. Kirby desires to assure Dr. Dewhurst of his great esteem.

This is to certify, that Mr. H. W. DEWHURST, hath attended my Lectures on the Theory and Practice of Surgery, during Four Courses, diligently in 1822, and the subsequent years.

B. C. BRODIE.

March 30, 1881.

When the Council of the King's College, London, advertised for Professors, I was ambitious to hold one of the Professorships in that excellent establishment, but was unsuccessful for want of interest and patronage. My surgical preceptor, Mr. Brodie, being on the Council, I addressed a letter to that gentleman, soliciting his support, when I was favoured with the following reply.

16, *Saville Row, August* 16, 1830.

My DEAR SIR,

I should have been happy to meet your wishes, but as a member of the Council of King's College, I have found it necessary to decline giving testimonials to any of the Candidates who offer themselves to us for the Professorships of that Institution.

Yours truly,

B. C. BRODIE.

To H. W. DEWHURST, ESQ.

Subsequently I applied to Mr. Brookes, but was again unfortunately too late, as the following letter testifies.

DEAR SIR,

Having already written a recommendation in favour of Mr. ———, I sincerely regret it is out of my power to accede to your request, which, had you applied earlier, I should have had much pleasure in granting.

I remain, with best wishes, yours truly,

JOSHUA BROOKES.*

Blenheim Street, July 26, 1831.

---

* See note to Mr. B.'s former letter, p. 300.

EXTRACT.

18, *Nelson Street, Commercial Road,*

16th *March,* 1831.

DEAR SIR,

I received the package you did me the honour to forward, and read its contents with much satisfaction, especially your very neat, comprehensive, and instructive Synopsis, " *A Dissertation on the Component Parts of the Animal Body.*" You have in these little tracts contributed essentially to those *most important* branches of human knowledge, *Anatomy, Physiology,* and *Animal Chemistry,* by placing the materials in a very simplified and very clear arrangement. In the modern rage for novelty, the task of arranging and clearing the science of obscurity and confusion is of the utmost importance, and requires more solid erudition and study than usually fall to the lot of fortunate and often visionary discoverers.

* * * * * * * * * * * * *

I should feel obliged by your enclosing per post your Phrenological card.

I am, Sir, most respectfully, your obedient servant,

J. HANCOCK, M. D.

———

Extract from a letter from Thomas Firth, Esq., Professor on Anatomy and Surgery.

48, *Clifton Street, Finsbury Square.*

DEAR SIR,

I thank you for your politeness in offering to send me the remarks which you are about to publish, * and beg to assure you that it will always afford me, not only pleasure but instruction in the perusal of any thing from the pen of so distinguished an Anatomist as yourself.

* * * I am, dear Sir, yours truly,

THOMAS FIRTH.

21st May, 1831.

To H. W. Dewhurst, Esq., Professor of Anatomy, &c. &c.

———

Extracted from the Scientific Gazette of June, 1831.

" Mr. Dewhurst, who must be very advantageously known to the readers of the *Scientific Gazette,* for his interesting and important communications to our Journal, is an active and indefatigable cultivator of science in one of its most *important,* if not its most *popular* departments. It has long been a matter of just complaint that science has been too much obscured by its technical terms ; this in a very peculiar manner applies to Anatomy. In one department of this subject, Mr. Dewhurst has successfully applied in lieu of an abstruse and unconnected nomenclature

———

* The Essay on the minute Anatomy and Physiology of the organs of Vision in Man and Animals.

terms more appropriate both for tuition and reference, because they are founded upon the distinguishing characteristics of the parts. Anatomical nomenclature is susceptible of yet further revision, and we think that the original powers of thinking which Mr. Dewhurst possesses, would enable him to carry a correcting hand with much advantage."

———

MUNDEN TERRACE, HAMMERSMITH.

DEAR SIR,

Last evening, after the termination of your Lectures, the following resolution was unanimously agreed to, and which I have much pleasure in transmitting to you.

I remain yours, &c.

JAMES CHARLES RUSSELL.

August 12, 1831.

J. REEVES, ESQ., in the Chair.

Resolved unanimously, that the thanks of this meeting be given to Professor Dewhurst, for his able and comprehensive course of Lectures on Human and Comparative Anatomy, which he has just concluded, at No. 10½, Great Titchfield Street, Cavendish Square.

———

LE DR. RUCCO,*
A Monsieur le Professeur H. W. Dewhurst.

MONSIEUR,

Je me fais un devoir et un plaisir, à la fois de vous remercier bien sincèrement du présent qu'il vous est plu de me faire de vos *productions scientifiques;* d'autant plus que je n'y avais aucun têtre à cet honneur : le procédé, de votre part, m'a engagé à les lire, et même à les étudier autant qu'elles m'ont inspiré de l'intérêt. D'abord, je dois vous feliciter de votre réforme relative à la *nomenclature* des sutures du crâne ; du choix de la méthode que vous avéz suivie aussi bien dans l'enseignement de différentes branches de la science de l'homme, et des autres animaux que vous professez, que dans les recherches que vous y avez consacrées pour leur avancement et finalement de votre zèle et de vos efforts pour les rendre populaires, et par là les faire concourir au bien public. Mais ce serait sortir des limites d'une lettre que vouloir

———

* From Dr. Rucco, formerly Professor of Physiology and Comparative Anatomy in the University of Pavia, General Inspector of Military and Civil Hospitals throughout the Empire of France and the Kingdom of Italy, and personal Physician to the Emperor Napoleon the First and the Empress Marie Louise, &c., and author of a valuable work on the " Science and Physiology of the Pulse."

relever d'autres points eminents, de vos ouvrages. Le fait est qu'en vous lisant, j'ai acquis une juste idée de votre mérite, et de vos talens. Je serai même plus qu'un jeune professeur de votre goût et de votre èsprit d'examen pourrait aller encore bien loin, surtout dans l'exuse de l'anatomie comparée (source inépuisable de connaisances utiles), si des circonstances favorables venaient à votre appui: il vous faudrait, pour cela, une chaire publique dans un de vos colleges avec des encouragemens qui pourraient vous mettre à même de suivre le développement de cette science vraiment utile dans les plus petits details, sait par des nouvelles recherches, sait à l'aide d'expériences plus directes.

En fin, si l'opinion favorable que j'entretiens de votre habilité dans l'enseignement soit de l'anatomie humaine, soit de l'anatomie comparée, pouvait se soustraire à l'empire des prejuges qui existent spécialement chez-vous contre le savoir quelque soit de savants étrangers, et, par le moyen vous être utile ; je 'n'hesiterais, point à répéter ce que j'avance aujourd'hui que les connaissances que vous avez acquises, et qui sont le sujet de vos ouvrages, vous rendent digne de votre élévation à la chaire d'anatomie humaine, ou comparee qui exige plus d'habilité de la part du professeur, ètant convaincu d'avance que vous sauriez justifier mon opinion par l'utilitè et l'importance de vos leçons, et de vos travaux.

J'ai l'honneur d'être,

Monsieur,

Votre très humble, et devoué serviteur,

J. RUCCO.

13, Howland Street, Fitzroy Square,
Le 27 Août, 1831.

LISSON GROVE.

This is to certify, that I have been personally acquainted with Henry William Dewhurst, Esq., Surgeon-Accoucheur, for several years, during which time he has evinced considerable talent as a Professor of Human, Veterinary, and Comparative Anatomy, and teacher of Midwifery.

I have much pleasure also in stating that I consider Mr. Dewhurst duly qualified to fill a public Professorship of either Human or Comparative Anatomy.

Witness my hand, this 31st of August, 1831.

SAMUEL NICE,* SURGEON.

---

* This gentleman was late one of the district Surgeons of the Board of Health in the Parish of Saint Mary-le-bone.

From Dr. F. Macartney, Physician to the Right Reverend Father in God, the Lord Bishop of Ferns.

Enniscorthy, September 4, 1831.

I have perused with great satisfaction and advantage, the various treatises of Mr. H. W. Dewhurst on Anatomical and other subjects; and from his correct and intimate acquaintance with the healthy structure and functions of the human body, from his superior abilities, unwearied industry, and great sagacity in investigating and detecting the morbid appearances that present themselves after death, and from his peculiar taste for the study of Comparative Anatomy, and the progress he has made in that department of science, I look upon him as eminently qualified to fill the situation of Professor of Anatomy in any British College. The urbanity of his manners, the ardent zeal he displays on all occasions to extend the boundaries of human knowledge, and his peculiarly eloquent and felicitous manner of imparting to others the information he has acquired, form additional recommendations that must render any competition with him one of no ordinary pretensions, and allowing of no inferior or deficient attainment.

F. MACARTNEY, M. D.
SURGEON, FERMANAGH REGIMENT.

———

DEPTFORD MECHANIC'S INSTITUTION, HIGH STREET.

DEAR SIR,

In reply to your note, I have much pleasure in stating that the various courses of Lectures on ANATOMY, PHRENOLOGY, and NATURAL HISTORY, delivered by you before the Members of this Institution, have given universal satisfaction; and I feel great pleasure in adding that I am requested by the Committee of Management to tender to you their thanks for the zeal and interest you have so kindly and uniformly displayed in promoting the welfare of this Institution.

I have the honour to remain,
My Dear Sir, yours faithfully,
J. RUSSELL, HON. Sec.

September 16, 1831.
H. W. Dewhurst, Esq.

———

7, HANOVER PLACE, REGENT'S PARK.

This is to certify, that having attended a Course of Lectures on Human and Comparative Anatomy, recently delivered by Professor Henry William Dewhurst, and from my intimate knowledge of the necessary acquirements for a public Lecturer, I have not the least hesitation in stating that Professor Dewhurst's discourses met my most unqualified approbation, inasmuch as he evinced considerable talent in his descriptions of the various subjects of his Lectures, he was peculiarly happy in his illustrations, at the same time clear and forcible in his delivery. His specimens, drawings, and preparations, prove him deserving of the highest credit, and worthy of public patronage. I therefore conceive him fully qualified to fill the situation of Professor of either Human or Comparative Anatomy in any College or University.

Witness my hand, this 26th day of September, 1831.

J. NASH, SURGEON-ACCOUCHEUR,
And Cupper to the Royal Ophthalmic Hospital.

MIDDLE TEMPLE, *September,* 1831.

I **Certify,** that I attended some time ago a course of Lectures delivered by Professor Dewhurst, upon Comparative Anatomy, Physiology, and Pathology; not being a medical man, I cannot give so decided an opinion, as had I the honour to belong to that most distinguished profession, but from the partial knowledge I possess of the above-named branches of science, I beg to state, I have formed a very superior estimation of Mr. Dewhurst's capabilities. His language was elegant and effective, his style chaste and forcible, and his delivery truly impressive in all respects: He struck me as a gentleman of talent, and highly fitted for a public Lecturer.

W. J. A. ABINGTON, A.B.
Of Trin. Coll. Cam., and of the Middle Temple, London.

HAGUE, *September 30,* 1831.

DEAR SIR,

In reply to your letter of the 21st, I feel much pleasure in adding my opinion to the numerous testimonials, which I am aware you have received, regarding your being properly qualified to practise the three branches of the medical profession. Assure yourself that it is no flattery on my part, when I affirm, I should feel quite confident in recommending you to any family who might feel disposed to consult you as their medical adviser. Accept my best wishes for your success in practice, and believe me to remain,

Yours ever sincerely,
HENRY F. W. VON HEYDELOFF, M.D.
Late Assistant Physician-Accoucheur to the Edinburgh Lying-In Institution.

Haag Street, No. 14.

P. S. I send this through the medium of a friend returning to England.

CAMDEN TOWN,
October 4, 1831.

MY DEAR SIR,

If my testimony can be of service to you, I have much pleasure in stating that since 1826 I have had numerous opportunities of becoming acquainted with your ability as a Lecturer, from an almost constant attendance on your Public and Private Lectures, during which I have derived much valuable information; I may observe that a perusal of your works has afforded me no inconsiderable share of instruction.

I remain, my Dear Friend, yours sincerely,
JAMES E. BROWN.

6, LEICESTER PLACE, LEICESTER SQUARE,
October 4, 1831.

DEAR SIR,

I regret that I am unable to send you the testimonial you request (at least so as to be deemed an official one, as chairman on the occasion alluded to) without the concurrence of the Committee of Management of which I am only an individual member.

I can therefore only express my own private sense of the pleasure I derived from your Lectures delivered in the Theatre of the Western Literary and Scientific Institution. I feel, however, in the same manner, no hesitation in stating, that they appeared to afford *very general satisfaction to the members at large.*

I remain, Dear Sir, yours respectfully,
JOHN POWER, M.D.

To H. W. DEWHURST, ESQ.

This is to Certify, that Mr. HENRY WILLIAM DEWHURST hath very diligently attended *Four Courses* of my LECTURES, DEMONSTRATIONS, and EXAMINATIONS, in the Sciences of HUMAN and COMPARATIVE ANATOMY, ZOOLOGY, and MENTAL PHILOSOPHY, during the years 1822 and 1823: during which period, he minutely and carefully dissected twice the Human Body, together with many of the lower animals, (amongst which I may mention the *horse, dog, sheep, lion, tiger, cat, tiger-cat,* and most of the *mouse tribe*), to my entire satisfaction. I may likewise state that his conduct was such, as to merit the regard of his fellow students, as well as my esteem; which I am happy to say continues up to the present period.

Witness my hand, this 13th day of November, 1831.

JOHN BROWNE, M. D.*
Bedford Street, Bedford Square.

---

* This gentleman was formerly Surgeon to the Right Rev. the Lord Bishop of Norwich, and also to His Grace the Duke of Leeds.

In the New Monthly Magazine for December, 1288, the Mechanics' Magazine, and Weekly Free Press, of about the same date, the Morning Herald of September .5, 1830, the Gazette of Health for the same year, and in the New Monthly Magazine of January 1832, the Maidstone Journal of October 22, and the Maidstone Gazette of October 22, and 29, 1833, copious Reports of my various Courses of Lectures will be found, and to which the reader is most respectfully referred.

From J. RENNIE, Esq. A.M. Professor of Zoology, King's College, London.

LEE, KENT, 29th February, 1832.

MR. RENNIE presents his compliments to Professor Dewhurst, and thanks him for the copy of his book, (the Lecture on the Architecture of the Human Body), and regrets that he has no connection with any Periodical in which he could give a notice of it, such as it certainly well merits.

To H. W. Dewhurst, Esq.
Professor of Zoology, Anatomy, &c.

This is to certify that I have known Professor HENRY WILLIAM DEWHURST, since October 1828, during which period, I have had numerous opportunities of becoming acquainted with his abilities as a Public and Private Teacher of the Sciences of ANATOMY, ZOOTOMY, NATURAL HISTORY, and PHRENOLOGY; also as a Medical Writer and General Practitioner; and it gives me much pleasure to state that I consider him to be a man of irreproachable reputation.

Witness my hand, this 20th day of March, 1832.

WILLIAM HUNT,*
SURGEON-DENTIST.

No. 2, Manchester Street, Manchester Square.

From David Mallock, Esq., A.M., author of a most beautiful Poem, entitled "The Immortality of the Soul."

April, 1832.

DURING the winter of 1830, H. W. DEWHURST, Esq., PROFESSOR OF ZOOLOGY, &c. &c., delivered before the Members of the Westminster Co-operative Institution a Course of Lectures on HUMAN and COMPARATIVE ANATOMY; and I have the greatest pleasure in stating that these Lectures gave unqualified satisfaction. Though the subject, especially to those who had little or no opportunity of previously becoming ac-

* I was honoured by this gentleman dedicating to me his ingenious and valuable little work on " *Diseases of the Teeth*."

314 APPENDIX.

quainted with Physiological Science, appeared on its first announcement to be barren and uninteresting, yet Professor Dewhurst, by laying aside all technicalities, and simply and beautifully illustrating every part of his subject, gave such variety and interest to every topic which he handled as to render the whole a source of much gratification and instruction, and to add another to the many proofs which Professor Dewhurst has given of his experience in his profession, and his ability as a public instructor.

Signed.　　　　　DAVID MALLOCK, A. M.

Hon. Secretary.

3, Cannon Row, Parliament Street, Westminster.

Extract of an unsolicited letter from the Rev.Mr.Robert Wrench, A.M., of 90, Sloane Street, Chelsea,

Monday Morning, June 17, 1833.

Dear Sir,

Your admirably interesting and entertaining "*Lectures on the Architecture of the Human Body*" have been the theme of my eulogium for some time, as well as recommendation to my friends and young neighbours; and you amply deserve this humble tribute of applause from an old member of the Church of England, now in his seventieth year.

* * * * * * * * * * * * *

Yours faithfully,

ROBERT WRENCH.*

This is to certify, that I have been acquainted with Professor HENRY WILLIAM DEWHURST, for the last *Eight Years*, during which period I have had numerous opportunities of ascertaining his abilities as a public Professor of Zoology, Popular Anatomy, and Phrenology, as well as the author of many truly Scientific Treatises, all of which reflect great credit on his talents. His moral reputation I consider to be irreproachable.

Witness my hand, this 20th day of July, 1833.

JOHN WOLSEY BAYFIELD.†

Walworth.

---

* This gentleman did me the honour to consult me in this letter, and, when it was written, I was unacquainted with him, either by name or otherwise.

† This gentleman was lately a Lieutenant in the Third Regiment of Dragoon Guards.

Maidstone, October 26, 1833.

DEAR SIR,

I have much pleasure in making known to you, that after the conclusion of your *Lectures* on the *Zoology* or *Natural History of Man*, Thomas Sponge, Esq., being still in the chair, it was on the motion of Mr. Heathorn, seconded by Mr. Wright, unanimously resolved as follows : " That the thanks of this meeting be presented to Professor Dewhurst for the interesting communications made in the course of his Lectures on the Zoology or Natural History of Man, and from which much gratification has been derived by all present."

In making this communication, receive also my best thanks for your kind attentions, and be assured of the esteem of,

Dear Sir, yours very truly,
SAMUEL SIMMONDS.

Mechanic's Library.
To Professor Dewhurst, London.

————

EXTRACT.

Worcester, November 24, 1833.

MY DEAR SIR,

I have been so much engaged as to be unable to write to you sooner. I have, however, now the pleasure to inform you that you have been elected a *corresponding Member of the Worcestershire Natural History Society*.

The council request me to convey to you their thanks for your donation to their library.

Believe me, Sir, yours very sincerely,
EDWIN LEES

To PROFESSOR DEWHURST.

316

316   APPENDIX.

PROFESSOR DEWHURST'S
POPULAR LECTURES
ON
NATURAL THEOLOGY.

" Astronomy, Anatomy, and Natural History, are the studies which present us with the most striking view of the two greatest attributes of the Supreme Being. The first of these fills the mind with the idea of his immensity, in the largeness, distance, and number of the heavenly bodies; the two latter astonish with his intelligence and art in the variety and delicacy of animal mechanism."—*Fontenelle.*

MANAGERS of Philosophical, Literary, and Scientific Institutions, Mechanics' Institutes, Private Societies, &c., desirous of having any of the under-mentioned Lectures delivered before their members, are requested to address a letter, *postage free*, to Professor Dewhurst, at his residence, which will meet immediate attention.

The preceding Testimonials are respectfully solicited perusal.

Pupils prepared privately for examination at the Universities, Royal Colleges of Physicians and Surgeons, Royal Veterinary College, Apothecaries' Hall, Army, Navy, and East India Medica Boards, &c.

Terms for Private and Public Lectures extremely moderate.

SYLLABUS OF THE LECTURES
ON THE
ANATOMY, PHYSIOLOGY, AND MECHANICAL
STRUCTURE OF THE HUMAN BODY,
WITH ITS CHEMICAL FORMATION.

LECTURE I.

A FAMILIAR introductory| discourse, comprising a brief but comprehensive general description of the various parts composing the human body compared with many of the lower animals—uses of the science of popular anatomy to mankind.

LECTURE II.

Description of the formation, structure, uses, and chemical composition of the skeletons of the various orders of animals,

LECTURE III.

The mechanical structure of the skeletons of man and quadrupeds—architecture of the human skull, in the fœtus and adult — Dewhurst's improved nomenclature of the cranial sutures — conformation of the skulls of various animals, and of man, in different nations—analogy between Collinge's spherical hinge and the hip-joint—arched form and elastic structure of the feet of man and quadrupeds.

LECTURE IV.

Formation and uses of the blood, chemical analysis—action of the air upon the venus blood—composition of air—changes the blood undergoes in the lungs—colour of this fluid in different animals—recent discovery of oil in the blood by Dr. Trail of Liverpool—History and discovery of its circulation by Dr. Harvey, in 1620.

The actual circulation exhibited after the Lecture, in the *Rana Temporaria*, or common frog, when they can be procured.

LECTURE V.

The structure and uses of the circulatory organs; viz. the heart, arteries, and veins, in all classes of animals—importance of a knowledge of the course of the principal blood-vessels to officers in the army, navy, and public at large.

LECTURE VI.

The anatomy and physiology of the organs of respiration, digestion, and nutrition—remarks on the various substances employed as food.

LECTURE VII.

Examination of the structure and uses of the organs of vision in man and the various classes and orders of animals, illustrated by an *actual dissection of the eyes of different quadrupeds*.

LECTURE VIII.

The anatomy and functions of the organs of sense; viz. hearing, smell, taste, and touch.

LECTURE XI.

The structure and functions of the brain and nervous system in man and animals; together with an explanation of the theory of the mental faculties, as ascertained by the ancient metaphysicians; also of the physiognomical system of Lavater, and the phrenological doctrines as promulgated by the late Drs. Gall and Spurzheim.

LECTURE X.

The uses and powers of the organs of locomotion in different orders of animals; *i. e.* the anatomy of the muscular system—strength of man and animals—advantages arising from gymnastic exercises—dancing, &c.

LECTURE XI.

The structure and physiology of the organs of procreation, in the human species, quadrupeds, birds, and reptiles of both sexes—reproduction of the species in animals and plants—duration of pregnancy in the human female, and the lower animals—Can protracted pregnancy ever occur? —importance of this subject in a medico-legal point of view —decision of this question—Conclusion of the Course.

The whole of the above Course forms a familiar but comprehensive view of the Anatomical, Physiological, and mechanical formation of the Human Body, together with a comparison of the structure of quadrupeds, birds, fishes, and reptiles, which cannot fail to prove highly interesting to the Philosopher, Medical Student, and Practitioner, as well as to the general Student of the Arts and Sciences; and (with the exception of Lecture XI.) there is nothing exhibited or expressed that would preclude the attendance of Ladies, whose company is respectfully solicited.

Professor Dewhurst begs to observe that the Lectures numbered II. III. IV. V. IX. XI. in the above Syllabus may be condensed, if required, into four Lectures. However, should it be considerable to enter more comprehensively into the subject, they should be delivered in the order above mentioned.

LECTURES

ON THE

ZOOLOGY OR THE NATURAL HISTORY

OF THE

VARIOUS RACES OF MANKIND.

IT is not a little remarkable that with the exception of Professor Lawrence, at the Royal College of Surgeons, in 1816, and myself at the different Literary and Scientific Institutions in London and the Provinces, no Public Lecturer has thought it worth while, or of sufficient importance, to deliver a Course of Public Instruction on the Zoology, or Natural History, and National Peculiarities, of the various races of mankind; in fact of the *Chef d'Ouvre* of the wondrous creative power of an Omnipotent God, respecting which the beautiful and well-known aphorism of the immortal Pope ought to be in constant remembrance,

" That the proper Study of Mankind is Man."

The following constitute the subject of this Course of Lectures.

LECTURE I.

Introductory observations—Origin of Man—Classification of Man—Form of Man—Anatomical Peculiarities—History of Man—Multiplication and Dispersion of the Human Race —External Configuration—Complexion, Progressive condition of Man—General Peculiarities of the Two Sexes— Mental Characteristics of Woman—Description of a Singalese Beauty—Mr. Sharon Turner's Observations—Temperament of the Constitution—Man enabled to live in any Climate—Food and Antipathies of Man—Conclusion.

LECTURE II.

The First Age or Infancy of Man—Peculiarities of the Early Stages of Human Existence—Second Age—Teething —Third Age—Childhood—Fourth Age—Youth—Stature of Man—Fifth Age—Manhood—Strength of Mankind— Differences in the Constitution, consonant with the peculiarities of Climate—Beauty of the Female Sex—Fertility of

Soil, in accordance with the increase of Human Species—
The Human Frame affected by peculiar changes of the
Atmosphere—Sixth Age of Man—Periods Fatal or Inju-
rious to the Health of Man—Hereditary Longevity—Reca-
pitulation—Last Age of Man, Decrepitude and Death—
Conclusion.

LECTURE III.

Influence of the Nervous System on the Constitution—Ef-
fects of Art in the Configuration and Features of the Human
Race—Professor Blumenbach's Classification—Circassians
and Georgians—Europeans—Scotchmen and Englishmen—
Jews—Egyptians, and Gypsies—Asiatics—Greenlanders—
Tartars—North and South Americans—Caribs—Ethiopians
—Negroes—their intellectual characters, sufficient alone to
prevent their being employed as slaves—Lawrence, Cuvier,
Hume, Gibbon, and others, refuted by facts—Hereditary
variety of the Human Configuration—The Last Age of
Man, or the End of the Existence of the Human Species—
Conclusion of the Course.

** Any of the above Lectures can be delivered separately,
without interfering with the arrangement of the subject of
any of the others, and at a few hours' notice.

These Discourses are familiarly illustrated by a most
extensive and splendid series of Preparations, Fœtal and
Adult Skeletons, Tables, Drawings, Coloured Engravings,
Diagrams, Skulls and Casts, Models, &c. &c.

LECTURES

ON

THE NATURAL HISTORY

OF THE

ORDER CETACEA

As soon as the illustrations to these interesting Lectures
on the Marine Zoology of the Polar Seas are completed,
Professor Dewhurst will duly announce them.

PROFESSOR DEWHURST'S ADDRESS TO THE AFFLICTED PUBLIC.

" Science should contribute to the comfort, health, and happiness of mankind."

PROFESSOR DEWHURST, F.W.N.S., Consulting Surgeon-Accoucheur, &c., most respectfully announces to his friends and the public, that he may be daily consulted (*at first by letter*), in all cases requiring medical or surgical assistance, particularly in all nervous complaints; chronic indigestion, and that of elderly people; spasms of the stomach and bowels; general relaxation, inflammations, and other affections of the lungs, producing consumption; loss of appetite; languor; tremor; palpitation; palsies; despondency; want of energy; chronic; head-ache; costiveness; worms, especially those which trouble children; hysteric diseases; female comlaints, connected with debility, and dropsical swe llings of the legs; restlessness, and frightful dreams; rheumatism, gout, &c., and most disorders affecting the human frame.

The sufferers from these diseases may, by early application to PROFESSOR DEWHURST, meet not only with speedy relief, but likewise an effectual cure. In the treatment of the numerous diseases which the human flesh is heir to, the various symptoms afflicting the patient are taken into consideration, and every case treated as its urgency may require. It being utterly impossible, notwithstanding the present reign of QUACKERY, that any one medicine should cure every disease. From the long experience PROFESSOR DEWHURST has enjoyed in the practice of the three branches of the profession, he has devoted himself to the study and cure of disease, and not to that of mere mercenary motives, the sole object of the unblushing empirics, who, to the disgrace of the country, infest particularly this metropolis, and after draining the pockets of their unfortunate victims, leave them in a miserable condition both of body and mind, and in the majority of cases, the constitution totally destroyed.

In offering himself to the afflicted public, as a practitioner of the healing art, PROFESSOR DEWHURST begs to observe that his method of cure is founded entirely upon his

Y

knowledge of the human body in its healthy condition, and the morbid changes which occur in its organization and the performance of the various functions which the Deity has assigned to it; and not upon any of the nonsensical theories which the *Empirics* of the present day impose upon the public.

Having been regularly educated in every branch of the Medical Profession (as his numerous Testimonials from some of the most illustrious men in Great Britain and Ireland will certify), and also actively engaged in the performance of the various duties of his profession for the last seventeen years, likewise as the author of upwards of *twenty* works on the most important branches of medical science, besides numerous papers on the nature and treatment of diseases in the different Medical Journals; and as a public Professor of Human and Comparative Anatomy, Midwifery, Zoology, &c.; he trusts his claims to public confidence are not solicited without some solid foundation.

DISEASES PECULIAR TO FEMALES.

" The maladies incidental to Females, destroy more than two-thirds of the Fair Sex; either from their complaints not being properly comprehended by their medical attendant; and in many cases from their own neglect in the earlier stages of disease." *Dr. Denmʼn.*

THE sedentary lives led by the majority of females in this country, not unfrequently produce either an obstruction of an important monthly function, the due performance of which is necessary not only to the health of the sufferer, but likewise to place the female in that condition which nature requires in the married state, to enable her to become a mother; or sometimes we find that nature has been too profuse, thus in both these cases we find a constitution in itself naturally delicate, severely debilitated, and unless it is attended to, on scientific principles, organic diseases are engendered, and the amiable sufferer is thus consigned to an early grave.

There is another series of diseases, which, if not of so dangerous a nature as the preceding, are as important, viz. "*Leucorrhœa,*" To sufferers from this affliction, but little need be said to point out the necessity of an early cure, for unless it is speedily checked, cancer and other diseases of the womb occur, which whether the patient be *married* or *single*, a state of mental and bodily suffering is induced, and which is only terminated by death closing the melancholy scene. By the method of cure pursued by PROFESSOR DEWHURST, the patient may confidently rely on being speedily restored to her original state of health, and that in the course of a very few days.

Having been frequently afflicted with inflammation of the lungs and other diseases of the chest, tending towards consumption, to which he had nearly fallen a victim, PROFESSOR DEWHURST is happy in being enabled to assert with confidence, particularly to the fair sex who are unfortunately so much afflicted, that he has not only *discovered a remedy*, which he has found beneficial in curing himself, but likewise numerous individuals have enjoyed its advantages. He feels also happy to state, that having published to the medical Profession the results of his discovery, they have uniformly and in the most public manner acknowledged its value ; consequently no disgraceful empiricism forms a part of his method of cure.

PROFESSOR DEWHURST may be consulted by patients in the remotest parts of the country, who can be treated successfully on transmitting a minute account of the symptoms causing their complaint, which will be immediately answered; this communication must contain a remittance for advice and medicine, the remedies for which will be forwarded to any part of the world, however distant.— No difficulty can occur, as the medicine will be securely packed, and carefully protected, and the patient regularly corresponded with until cured.

₊ *All unpaid letters are invariably refused.*

LIST OF PROFESSOR DEWHURST'S WORKS.

With the Opinions of some of the Reviewers, &c.

1. A LECTURE illustrative of the ARCHITECTURE of the HUMAN BODY, intended for the Rising Generation, to which is added a complete MANUAL OF ANIMAL CHEMISTRY.—*Sixth Edition.*

" This little work comprises many sensible and interesting observations, delivered in language easily to be comprehended by ordinary capacities. The copy before us is announced as belonging to the fifth edition, so that the circulation of this Lecture must have been extensive."—*Imperial Magazine,* May, 1832.

" What delights me most with this humble production is the pious and benevolent feeling in which it seems to be composed ; the author invariably tracing the wonderful mechanism of our corporeal frame to the *forethought* and *design* of an Omnipotent Creator. There is such a simplicity in the language, also, that it can be read with edification by even the most unlearned."—*Chambers' Edinburgh Journal, Aug.* 25, 1832.

This little work has met the approbation of General Viney, Dr. Birkbeck, Sir Astley Cooper, Rev. Dr. Drew, Rev. Mr. Long, and other eminent characters.

2. PRACTICAL OBSERVATIONS on the new System of WARMING DWELLING-HOUSES, Cathedrals, Churches, Theatres, and other Public Buildings with HOT WATER ; together with a description of the dangerous and uncertain effects produced by the employment of Heated Air, &c. &c.—This work obtained Major-General Viney's Prize of Twenty Guineas, in January, 1832, for which it was written.

" This contains much useful information, for all concerned in the above subject."—*Dr. Johnson's Medico-Chirurgical Review,* 1832.

" A highly interesting little work."—*Dr. Ryan's London Medical and Surgical Journal, March,* 1832.

" From the experience displayed by the author, we perfectly agree with his opinion, that an equal distribution of heat would preserve health, as also prevent and eradicate disease. To such of our readers as feel interested in this subject, we recommend an attentive perusal of this pamphlet, premising that it is intended chiefly for the general reader."—*National Omnibus,* Feb. 17, 1832.

" It is highly creditable to the abilities of the author, whether we view him as a philosopher, or as an intelligent medical practitioner."—*Weekly Visitor.* Feb. 26th, 1832.

" This is a truly valuable little Treatise. Every architect, engineer, builder, and invalid, should read it."—*Edinburgh Journal,* 1833.

" We have read this little book with great pleasure, and willingly give our tribute of praise to the lucid exposition of the new theory of warming houses by heated fluids in pipes, and the practical information so necessary to procure its general adoption. This essay gained the prize of twenty guineas, given by General Viney, for the best System of Warming Public Buildings, &c., and reflects high credit upon the author, of whose talents as a Lecturer upon Anatomical and Zoological Subjects we have before had occasion to speak. The extensive circulation of this little Essay, of which the price is hardly more than nominal, will probably save the nation many millions a year in the article of fuel. We have received some other of the Professor's works, but have not yet had time to read them with the attention they seem to deserve."—*Maidstone Gazette*, Nov. 12, 1833.

See also the "*Worcester Herald*" *of December* 28, 1833.

3. A LECTURE on the STUDY OF NATURAL HISTORY, adapted for the perusal of Youth.

"We cannot do better than strongly recommend to every parent to place this little book in the hands of their children; it will give them, in a clear and concise manner, an idea of the wonderful works of the Creation, and lead them pleasingly on to the study of the more complex works. We only regret that the highly talented author has not entered more largely into the subject. The price, however, would not admit of our desires."—*Weekly Visitor*, Feb. 1832.

4. PORTABLE BATH. "Professor Dewhurst has constructed an ingenius portable Vapour and Warm Air Bath, possessing all the advantages of Captain Jekyll's, at not one-sixth the cost. When taken to pieces, and put into the case, it is about the size of a writing desk. For travelling invalids, in particular, it is invaluable; vessels carrying passengers should not be without one of these useful machines."—*Weekly Visitor*, *March*, 1832.

5. A DISSERTATION on the COMPONENT PARTS of AN ANIMAL BODY; being a Lecture introductory to the Study of Human, Comparative, and Physiological Anatomy. *Fourth Edition.*

"The first edition of this introductory discourse was published in 1827, and obtained, as it deserved, the unqualified approbation of the most eminent characters in the medical profession; and received a similar well-earned applause at the various literary and scientific institutions, in and near the Metropolis, before whose members the talented author delivered his popular courses on Human and Comparative Anatomy.

"At the request of many of his friends, and at the call of the public, the learned author has republished his Dissertation, and in his new Edition has added great portions of new matter, which cannot fail to be highly interesting and instructive to the medical student and general reader. He has endeavoured, on this occasion, to avoid, as much as possible, the introduction of uncouth technical terms, which tend to prevent their general cultivation, are often perplexing to the pupil, and impossible to be understood by the public at large. In this Professor Dewhurst has shown the judgment of a superior and intelligent mind.

"We doubt not that the perusal of this Dissertation will be the means of making medical and veterinary students very much more attentive to the study of Anatomy than is generally done; and afford to the general reader an opportunity of obtaining possession of a small portion of information, and stimulating his zeal for further acquisition, which combined with a little more gained by actual instruction by a proper teacher (as for instance Mr. Dewhurst), may occasionally be the means of saving a fellow-creature's life, by compressing a large artery, thus checking a fatal hemorrhage, merely from his understanding the course of a blood-vessel.

"Into a critical analysis of this excellent dissertation we shall not again enter, having fully displayed the merits of the first edition in a former number. We have only to cordially recommend it to the attentive notice of the numeous professional and general readers of this extensively circulating

Journal."—*Gazette of Practical Medicine, June,* 1831.—Edited by the late Dr. Reece.

" The design of this little work is to introduce the medical student and general reader to a knowledge of animal structure, and of the elements into which the various substances which enter into its composition may be resolved by chemical analysis. This little volume displays considerable research and diligence, the materials having been collected from various sources, whilst the reference to the original and more voluminous works enable the reader to obtain fuller details, should they be desired. Its neatness and portability, in addition to the valuable scientific instruction which it conveys, will we doubt not render it acceptable to the public and useful to the author, whose zeal as a popular teacher at various scientific institutions in London and its environs is favourably known and appreciated."—*Athenæum.*

" A valuable little work and highly creditable to the author."—*Weekly Free Press.*

6. A DICTIONARY of PRACTICAL ANATOMY and PHYSIOLOGY. Part I., price 5s. 6d. Callow and Wilson.

*** This is the only work of the kind in any language, and will be found of great utility both to the student and practitioner.—*Extract of a letter from Dr. Ryan to the Author.*

7. A GUIDE to the STUDY of MENTAL PHILOSOPHY, according to the Phrenological Doctrines of Drs. Gall and Spurzheim, with Observations on the Formation and Varieties of the CRANIUM in MAN and ANIMALS ; to which is added, a Description of the New Method of Dissecting the Human Brain. *Out of print. A New Edition Preparing.*

8. A DIAGRAM of the FACULTIES of the HUMAN MIND, illustrating the THEORIES of the ANCIENT METAPHYSICIANS, designed by Mr. Dewhurst, price 6d.

9. A DIAGRAM of the MENTAL POWERS of MAN, in accordance with Dr. Spurzheim's Classification of the FACULTIES of the HUMAN MIND ; being in fact a beautiful Map of the Phrenological Organs, price 1s.

10. A SYNOPTICAL TABLE of an IMPROVED SYSTEM of NOMENCLATURE for the STRUCTURES of the CRANIUM. *Fifth Edition,* price 1s.

This system has met not only universal approbation from some of the most eminent anatomists in Great Britain, but has been introduced in the medical schools of Dublin, by Drs. Jacob, Houston, and Kirby, in lieu of those in common use, by Professor Monro, M. D. in the University of Edinburgh ; as also by Dr. Craigie, in the Encyclopedia Britannica.

11. A LECTURE introductory to the STUDY of PATHOLOGY and MORBID ANATOMY, price 6d. Originally inserted in the Gazette of Health. *Out of print.*

12. An ESSAY on the MINUTE ANATOMY and PHYSIOLOGY of the ORGANS of VISION in MAN and ANIMALS, dedicated to G. J. Guthrie, Esq. M. D. F. R. S. Professor of Anatomy to the Royal College of Surgeons. Price 8s.

13. SYNOPTICAL TABLES of the ANATOMY and PHYSIOLOGY of the HUMAN BONES, with plates, folio. Darton and Son, Holborn Hill. Dedicated to Dr. Granville, F. R S. &c. Price 15s.

14. A TABULAR VIEW of the ANATOMY of the HUMAN MUSCLES, with coloured plates, folio, price 15s. 6d. Darton and Son.

" These Works reflect great credit on the Anatomical abilities of their learned author."—*Medical Journal.*

15. OBSERVATIONS on the COMPARATIVE ANATOMY of the MUS MUSCULUS, or Common Mouse, dedicated by permission to His Royal Highness the Duke of Sussex. Price 1s.

16. OBSERVATIONS on the PROBABLE CAUSES of RABIES or MADNESS in the DOG and other Domestic Animals. Read at the London Veterinary Medical Society, on Wednesday Evening, Oct. 6, 1830. Price 1s.

17. PRACTICAL REMARKS on the Inutility of the HYDROSTATIC TEST in the DETECTION of INFANTICIDE, &c. &c. price 1s. *Out of print.*

18. CRITICAL REMARKS on the Impediments, Defects, and Abuses, in the present system of Medical Education, with Suggestions for their correction and removal.

" Were the system of study pointed out by you, followed by medical students in this country, it would *then be an honour to belong to the profession.*" —*Extract of a letter from Sir Astley Cooper to the Author.*

A new and enlarged edition of this work will shortly be published, dedicated to the Right Honourable the Lord Chancellor.

19. An ESSAY on the ZOOLOGY and NATURAL HISTORY of MAN. Price 6d. Houlston and Sons.

" An invaluable little treatise, on this shamefully neglected branch of zoology ; for what can be more interesting to Man, than a history of his own species ?"—*Scotsman.*

20. The NATURAL HISTORY of the ORDER CETACEA, and of the OCEANIC INHABITANTS of the ARCTIC REGIONS, in one octavo volume, illustrated with numerous lithographic and wood engravings. Price One Guinea.

" SIR ARTHUR DE CAPELL BROOKE considers that the work, (the price of which is extremely moderate), will be an interesting addition to any library."—*May 2, 1833.*

Opinion of SHARON TURNER, ESQ. F.S.A.

EXTRACT.

Red Lion Square, 26th Nov. 1833,

DEAR SIR,

I regret exceedingly that I should not have been at home when you called on Saturday.

My father desires me to return your book with many thanks. It has gratified him, and he will be glad to see it published. He thinks it will be *the best work on Whales we have.*

* * * * * * *

Believe me, with every sentiment of respect and esteem,
Sincerely yours,
W. TURNER.

PROFESSOR DEWHURST.

In the volumes of the Mechanic's Magazine, and London Mcehanics' Register, will be found many of Professor Dewhurst's Mechanical inventions and scientific observations.

21. A BIOGRAPHICAL MEMOIR of the late JOSHUA BROOKES, Esq. F.R.S., F.L.S., &c. &c.   Published in the National Portrait Gallery, by Messrs. Fisher, Son, and Jackson, Newgate Street, with an excellent Portrait, by T. Phillips, Esq. R.A.

" The Biographical Memoir of Mr. Brookes is unusually full and interesting "—*Court Journal, December* 14, 1833.

22. The POETICAL WORKS of JOHN MILTON, A. M., to which is prefixed a Biographical Sketch of the Author, by H. W. DEWHURST, Esq. Magnet Edition.—Price 4s. 6d.

" This beautiful and cheap edition of Milton's Poetical Works, is now complete.   The type is beautifully distinct and clear—the paper of excellent quality—and the text, so far as we have been able to examine it, carefully correct.   A head of the immortal bard forms the frontispice to the work, and is followed by a biographical sketch, which, though necessarily brief, contains all that is known of his eventful history.   The editor is Mr. Dewhurst, the Professor of Anatomy, to whom it is highly creditable."—*Morning Advertiser, July* 26, 1833.

NEARLY READY FOR THE PRESS.

1. A DISSERTATION on the ANATOMY, PHYSOLOGY, and PATHOLOGY of the HORSE'S FOOT, with splendid coloured engravings and wood-cuts.

2. A LECTURE INTRODUCTORY to the STUDY of PHRENOLOGY, with two engravings.

5, & 6. FAMILIAR TREATISES on DIET, and CONSUMPTION of the LUNGS.

7. A PHYSIOLOGICAL HISTORY of MAN, tracing his gradual progress through all the various stages of Human Existence.

8. A GRAMMAR of PHRENOLOGY, with an illustrative engraving of the Animal Propensities, Moral Sentiments, and Intellectual Faculties of Man,

9. A HISTORY of HUMAN, VETERINARY, and COMPARATIVE ANATOMY,

10. A TREATISE on the MINUTE ANATOMY of the BRAIN, and NERVOUS SYSTEM.

9. A complete Course of Demonstrations on the MINUTE ANATOMY and PHYSIOLOGY of the BONES composing the HUMAN SKELETON, with Observations on their Diseases.

10. ELEMENTS of HEALTH, comprising Practical and Familiar Observations on Temperance, Corpulence, Bathing, Clothing, Sleep, &c.

THE HORSE'S BRAIN.

Mr. Dewhurst begs to inform the Members of the Medical and Veterinary Professions, that he is preparing for Publication a series of four Wax Models of the Horse's Brain, modelled by an eminent artist, from dissections made expressly for this purpose by himself.

329

FOR BEAUTIFYING THE SKIN AND COMPLEXION.

ROWLAND'S KALYDOR.

As a preparation for the Skin is in preference to all others, selected by the ladies as an indispensable toilet requisite, sustainer of a fine complexion and conservator of female beauty in all climates, and during all stages in the progression of life from youth to age. It has already become a favourite and indispensably appreciated article with female Rank, honoured with the AUGUST PATRONAGE of the ROYAL FAMILY of GREAT BRITAIN ; Her Majesty the QUEEN OF THE FRENCH ; the PRINCE and PRINCESS ESTERHAZY ; and the most distinguished Nobility, &c., in all civilized nations, and is zealously recommended by the most eminent of the faculty.

The ingredients of Rowland's solely genuine Kalydor are extracted from the most Beautiful Exotics, are of the mildest nature, WARRANTED PERFECTLY INNOCENT, yet powerfully efficacious, as a thorough cleanser of the skin ; it eradicates FRECKLES, PIMPLES, SPOTS, REDNESS, and all Cutaneous Eruptions, from whatever cause originating ; and transforms into radiant brilliancy the most SALLOW COMPLEXION.

By persevering in the use of the Kalydor, it gradually produces a clear and soft skin, smooth as velvet, actually realizing a delicate White NECK, HAND, and ARM ; and a healthy and juvenile bloom will in a short time be infallibiy elicited, whilst its constant application well tend to promote the free exercise of those important functions of the skin, which are of the utmost importance for the preservation of a BEAUTIFUL COMPLEXION, averting the characteristics of age, even to a remote period of human life.

GENTLEMEN after SHAVING will find it allay the irritating and smarting pain, and render the skin smooth and pleasant.

Price 4s. 6d. and 8s. 6d. per bottle, duty included.

₊ To prevent imposition the Name and Address of the Proprietors, as under, is ENGRAVED ON THE GOVERNMENT STAMP affixed over the cork of each Bottle.

"A. ROWLAND and SON, 29, Hatton Garden."

The following Testimonials of its extraordinary good effects have been selected from an immense number. Many from the Nobility MAY BE SEEN AT THE PROPRIETORS.

To Messrs. Rowland and Son.

Gentlemen,—I with pleasure acknowledge the singular benefit I have received from your excellent Kalydor. My face which has been subject to Inflammation and Eruption for years, is now restored ; and my friends to whom I recommended it, give it their decided approbation. Please to send six bottles per bearer of this note.

I remain, Gentlemen,
Your obedient Servant,

Bangor, Sept. 17, 1827.　　　　　　　　　　　A. H. S.

Y

Printed in the United States
By Bookmasters